IRINA RODICA RABEJA

WRITINGS I

DIGITAL IMAGE COMPRESSION   nD PRINTING
INTERNET OF THINGS   UNEMPLOYMENT SOLUTION
ENVIRONMENT POLLUTION   ASTRONOMY CONCEPTS

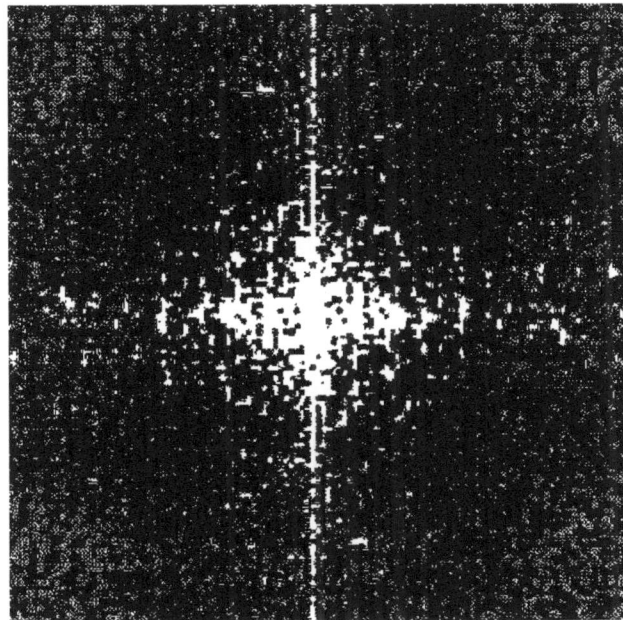

ENGINEERING   SCIENCE

IRINA RODICA RABEJA

WRITINGS I

DIGITAL IMAGE COMPRESSION   nD PRINTING
INTERNET OF THINGS   UNEMPLOYMENT SOLUTION
ENVIRONMENT POLLUTION   ASTRONOMY CONCEPTS

ENGINEERING   SCIENCE

National Library of Australia Cataloguing-in-Publication entry

Creator: Rabeja, Irina Rodica Olga, author.

Title: Writings I / Irina Rodica Rabeja.

ISBN: 9780977509850 (paperback)

Subjects: Engineering. Science.

Dewey Number: 620

Publisher Irina Rabeja
Sydney Australia
2025

CONTENT

DIGITAL IMAGE COMPRESSION TECHNIQUE

DIGITAL IMAGE

The Internet is the largest computer network in the world with millions users world wide giving possibility to people to access large volumes of data and exchange ideas fast.

Displaying photographs or images on a computer screen implies that the computer files of the photographs or images are transmitted via Internet through phone and cable lines or wireless and assembled by the browser software like Internet Explorer, Safari, Firefox, Netscape Navigator into an image on the computer screen.

The photographs or images have complex computer files that take more time to transmit than the words written on the computer monitor. The *file size* and its *transmission time* are key aspects of understanding how to make and use images on computers. The transmission time of an image or photo depends on the characteristic of the Internet called *bit rate*, which is the number of bits, electric impulses that are conveyed/processed per unit of time.

The computer images are made up of hundreds to millions of small squares called *pixels*.

A computer image is shown in FIG1.

FIG 1

The process of transforming an image or a photograph in a bunch of pixels and of giving information about the pixels - like their position and degree of grey or colour - in a digital form in a computer file is called *digitising*.

By digitising, the images or photographs are converted into *digital images* or *digital photographs*.

To appear on the computer screen and to be transmitted over the Internet a photograph or an image must be converted in its digital form.

There are several methods of digitising images/photographs:

- The main means for digitising are the devices called *scanners*. The scanner technology allows the conversion of static full-colour photos into digital pictures.

- The many commercially available *cameras* produce digital images. All digital cameras provide the hardware necessary to transfer images into a personal computer for easy transmission.

- Some modalities like *radiography* for example, provide completely digitised images, lending themselves to easy transmission.

Digital images have many advantages over standard static images because they can be manipulated with software for zooming/examining, can be rotated/encrypted, can be attached with audio/text files.

In digital imaging, a *pixel* or *picture element* is the smallest item of information of an image.

The word "pixel" is based on a contraction of two words, combination of the word "picture" contracted as *pix* and the word "element" contracted as *el*.

Pixels are normally arranged in a 2-dimensional grid and are often represented using dots, squares or rectangles. Each pixel is a sample of an original image and more samples provide more-accurate representations of the original.

The intensity of each pixel is variable, function of the level of grey of that pixel in a black and white image. A photograph, film or picture that is black and white or in black and white, contains only black, white and grey colours. In a coloured image each pixel has typically three or four components such as red, green and blue or cyan, magenta, yellow and black.

The concept "picture element" dates from the earliest days of television, for example as "Bildpunkt" a German word for "picture point" in the 1888 year German patent of Paul Nipkow.

The earliest publication of the term "picture element" itself was in Wireless World magazine in 1927, but it had been used as early as 1911 year in U.S. patents.

"Pix" had its first appearance in 1932 year in a Variety magazine headline, as an abbreviation for the word "pictures" in reference to movies and by 1938 year it was being used in reference to still pictures by photojournalists. The word "pixel" first published in 1965 by Frederic C. Billingsley of JPL Pasadena Ca. describes the picture elements of video images from space probes to the Moon and Mars. Some authors explain "pixel" as picture cell in 1972 year.

The more pixels used to represent an image, the closer their totality is to original.

The number of pixels in an image is called *resolution*. So *pixels* are used as *unit of measure* for resolution, such as: 3000 pixels per inch, 400 pixels per line or spaced 15 pixels apart.

For printer devices, *dots per inch dpi* is a measure of the printer's density of dot (ink droplet) placement. High *dpi* numbers do not mean high resolution.

So the measures *dots per inch dpi* and *pixels per inch ppi* have distinct meanings.

Megapixel MP (1 million pixels) is a measure used not only for the number of pixels in an image, but also to express the number of *display elements* of digital displays or the number of *image sensor elements* of digital cameras, picture-detecting rather than picture-producing elements.

The neologism *sensel* is sometimes used to describe the elements of a digital camera's sensor.

The pixels, or colour samples, that form a digitised image can or can not be in one-to-one correspondence with the computer screen pixels, that depending on the computer features.

In computing, an image composed of pixels is known as a *bitmapped* image or a *raster* image.

The word "raster" originates from television scanning patterns, meaning a rectangular pattern of parallel scanning lines followed by the electron beam on a television screen or computer monitor.

For convenience, pixels are normally arranged in a regular two-dimensional grid. By using this arrangement, many common operations can be implemented by uniformly applying the same operation to each pixel independently.

If the image is 300 pixels wide and 500 pixels high, it is said that it has a *size* of 300x500 pixels, in total 150,000 pixels. Each of the 150,000 pixels has a degree of grey or colour assigned to it.

The eye/brain system sees the pixel squares as single image of a real thing even though we really look at 150,000 tiny squares.

The degree of grey or colour of each pixel called *colour depth* is expressed by a binary number.

The binary number is expressed as a number of *bytes*, a byte being basically equal with eight *bits*.

So the colour depth of each pixel is expressed by a number of *bits per pixel* bpp.

Larger the binary number, larger bpp, more distinct shades of grey or colour are represented by the respective pixel, more details will have the image.

Formula for the calculation of the total number of shades done with a certain bpp is: 2^{bpp}

The simplest pixel representation is a black and white monochrome image in which one bit represents one pixel. Monochrome monitors as Cathode Ray Tubes CRTs use white, green or amber phosphors as a single colour over a grey/black screen background. Some monitors have the ability to vary the brightness of individual pixels, creating illusion of depth and color, like black-and-white television.

A 1 bpp image uses 1 bit or 0 bit for each pixel, so each pixel can be either *on* (white) or *off* (black):

0   1

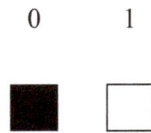

Each additional bit doubles the number of shades available, adding greys.

A 2 bpp image uses 2 bits for each pixel and the image has 4 shades ($4=2^2$).

A 3 bpp image uses 3 bits for each pixel and the image has 8 shades ($8=2^3$).

A 8 bpp image uses 8 bits (1 byte) for each pixel and the image has 256 shades ($256=2^8$):

00000000   ...   10000000   ...   11111111

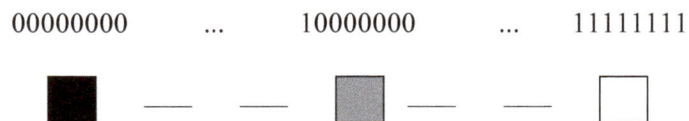

For 15 or more bits per pixel, the colour depth is normally the sum of the bits allocated to each of the red, green and blue RGB components. *Highcolour*, usually meaning 16 bpp, normally has five bits for red and blue, and six bits for green, as the human eye is more sensitive to errors in green than in the other two primary colours. For applications involving transparency, the 16 bits may be divided into five bits for each red, green and blue with one bit left for transparency.

Truecolour image uses 24 bpp (3 bytes for each pixel) corresponding to the primary colours red, green, blue and the image has 16.7 million colours ($2^{24}=16,777,216$):

RED	00000000	11111111	00000000	00000000	11111111	11111111	00000000	11111111
GREEN	00000000	00000000	11111111	00000000	00000000	11111111	11111111	11111111
BLUE	00000000	00000000	00000000	11111111	11111111	00000000	11111111	11111111

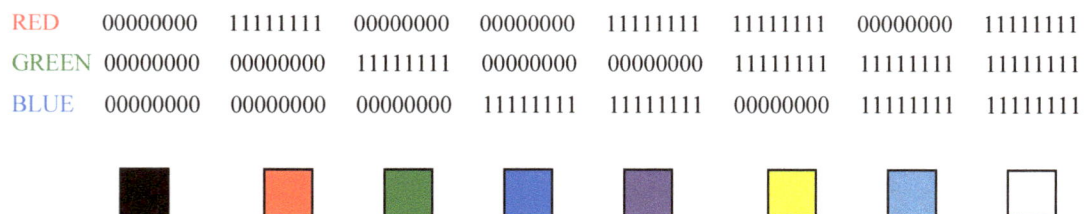

A 32-bit colour depth (4294 million colours) is available also on some systems meaning that each 24-bit pixel has extra 8 bits to describe its opacity (for purposes of combining with another image).

The *image file size* depends on the total number of pixels of image and on the number of bits per pixel bpp. Greater the total number of pixels, greater image resolution and larger is the file size and the time of transmission of image. Greater the number of bits per pixel bpp, greater image details and larger is the file size and the transmission time of image.

The computer file of a digital image is similar to a spread-sheet with rows and columns that stores for each pixel its colour depth. The colour depth of each pixel is the number of bpp allocated to each pixel and is generated by the image sensor of a device called *scanner*.

There are more types of scanners: drum scanners, flatbed scanners, hand-held scanners, film scanners, digital camera scanners, research scanners, document scanners, book scanners.

The scanners can be considered the successors of the early *telephotography* input devices, consisting of a rotating drum with a single photo-detector at a standard speed of 60 or 120 rpm with later models up to 240 rpm. They sent a linear, analogue, amplitude modulated signal through standard telephone voice lines to receptors, which synchronously print the proportional intensity on special paper.

Colour photos were sent as three separated red-green-blue RGB colour filtered images consecutively, but only for special events due to the transmission costs.

The scanners read data, process it to correct for different exposure conditions and sent to their host computer using one of the following physical interfaces:

Parallel Port, General Purpose Interface Bus GPIB, Small Computer System Interface SCSI, Universal Serial Bus USB, FireWire, Proprietary Interface.

The *first qualifying* parameter for a scanner is the *Colour Depth* that varies depending on the scanning array characteristics, but is usually at least 24 bits. High quality models have 48 bits or more colour depth. The *second qualifying* parameter for a scanner is its *Resolution*, measured in *pixels per inch* ppi, sometimes more accurately referred to as *samples per inch* spi, the number of individual samples that are taken in the space of one linear inch. In conformity to a formula, the size of the file created increases with the square of the resolution; doubling the resolution quadruples the file size.

A resolution must be chosen within the capabilities of the equipment, must preserve sufficient detail and must not produce a file of excessive size. The file size can be reduced for a given resolution by using "lossy" compression methods such as JPEG, at some cost in quality.

If it is required the best quality should be used "lossless" compression.

The *third qualifying* parameter for a scanner is its *Density Range*. A high density range means that the scanner is able to reproduce shadow details and brightness details in one scan.

The scanned result is an RGB image. Some scanners compress and clean up the image using embedded firmware before it is transferred to a computer's memory.

Once on the computer, the image can be processed with a raster graphics program, such as Photoshop or GIMP and saved on a storage device. Images are usually stored on the hard disk. Pictures are normally stored in image *file formats* such as uncompressed Bitmap, "non-lossy/lossless" compressed TIFF and PNG and "lossy" compressed JPEG.

A *file format* is a standard way that information is encoded for storage in a computer file. Documents are best stored in TIFF or PDF file format; JPEG is particularly unsuitable for text.

TIF and PSD image file formats have information to reproduce very well the image. However, they contain far too much information, far too many pixels, far too many colours, are far too dense in terms of pixels per inch.

The digital photographs or the digital images are described by the number of kilobytes kBs of information they contain. More kilobytes, longer time to transmit them via telephone/cable lines.

Minimizing the number of kilobytes without affecting the quality of the image on computer screen would be a choice for an optimal transmission.

Practically, minimizing the number of kilobytes can be done by reducing the size of image, by reducing the density of pixels and by reducing the number of colours, but each of those reductions affects the image quality.

-The smaller the image's width, height and surface in pixels, the fewer numbers of pixels to transmit and the faster the transmission time

-The smaller the number of pixels per inch, the fewer pixels to make up the image and the faster the transmission time

-The smaller the number of colours that the image is made from, the faster the transmission time.

The most imaging software reduces the number of colours automatically resulting in the use of fewer colours and thus lowers the number of kilobytes and fasters the transmission time. However, the image is also of lower quality and may lose sharpness [blur] or pixelate [patches of jagged edges, large areas of the same colour].

Usually when it is said that an image or photograph is prepared for transmission via Internet, that image or photograph is optimised not only by setting the height, width and ppi, but also by changing the image file format into the format called JPEG or JPG, which is designed to shorten transmission time through the Internet.

COMPRESSION TECHNIQUE

The use of digital computers in printing, publishing, video production and dissemination like broadcast television standards, tele-video conferencing, image sensors, remote sensing, satellite imagery for weather, document imaging, medical imaging, facsimile transmission FAX, control of remotely piloted vehicles is called *multimedia computing*.

In the multimedia computing there are generated images, which are light intensity function, combinations of dark and light of different levels of intensity in the 2D-space.

A continuous image, usually a photograph, can be approximated by a number of equally spaced samples, discrete quantities arranged in the form of an array MxN.

For practically very many samples, each element of the array is referred to as an *image element*, *picture element*, *pixel* or *pel*.

In computing, an image composed of pixels is known as *bitmapped* image or *raster* image, where for convenience pixels are arranged in a regular two-dimensional grid, in a "cartesian coordinate system" and so each pixel is a function of its pair of numerical coordinates xy.

Cartesian coordinate system is a coordinate system that specifies each point uniquely in a plane by a pair of numerical coordinates, which are the distances to the point from two fixed perpendicular directed lines, measured in the same unit of length.

Appears that the image has the function $I = f(x,y)$.

The image composed by pixels can be represented by a function of samples:

$$
f(x,y) = \begin{bmatrix}
f(0,N\text{-}1) & f(1,N\text{-}1) & f(2,N\text{-}1) & f(3,N\text{-}1) & f(4,N\text{-}1) & f(5,N\text{-}1) & f(6,N\text{-}1) & f(7,N\text{-}1) & \ldots & f(M\text{-}1,N\text{-}1) \\
f(0,N\text{-}2) & f(1,N\text{-}2) & f(2,N\text{-}2) & f(3,N\text{-}2) & f(4,N\text{-}2) & f(5,N\text{-}2) & f(6,N\text{-}2) & f(7,N\text{-}2) & \ldots & f(M\text{-}1,N\text{-}2) \\
\ldots & \ldots & \ldots & \ldots & \ldots & \ldots & \ldots & \ldots & & \ldots \\
f(0,5) & f(1,5) & f(2,5) & f(3,5) & f(4,5) & f(5,5) & f(6,5) & f(7,5) & \ldots & f(M\text{-}1,5) \\
f(0,4) & f(1,4) & f(2,4) & f(3,4) & f(4,4) & f(5,4) & f(6,4) & f(7,4) & \ldots & f(M\text{-}1,4) \\
f(0,3) & f(1,3) & f(2,3) & f(3,3) & f(4,3) & f(5,3) & f(6,3) & f(7,3) & \ldots & f(M\text{-}1,3) \\
f(0,2) & f(1,2) & f(2,2) & f(3,2) & f(4,2) & f(5,2) & f(6,2) & f(7,2) & \ldots & f(M\text{-}1,2) \\
f(0,1) & f(1,1) & f(2,1) & f(3,1) & f(4,1) & f(5,1) & f(6,1) & f(7,1) & \ldots & f(M\text{-}1,1) \\
f(0,0) & f(1,0) & f(2,0) & f(3,0) & f(4,0) & f(5,0) & f(6,0) & f(7,0) & \ldots & f(M\text{-}1,0)
\end{bmatrix}
$$

The right hand side of the equation represents what is commonly called a digital image sampling:

$f(x,y)$ becomes $\sum_{y=0}^{N-1} \sum_{x=0}^{M-1} f(x,y)$

By using this arrangement, many common operations can be implemented by uniformly applying the same operation to each sample/pixel independently.

In computers the images are used in their digital form, what means that the information for each pixel is represented binary by the number of bpp attached to it.

It is here to say that the image can be represented spatially in 3D, by mapping it in 2D with bpp as the third dimension.

A clear distinction must be made between data and information. They are not synonymous.

Data are the means by which information is conveyed. Various amounts of data may be used to represent the same amount of information, so a part of data either provides no relevant information or simply restates what already is known, conditions called *data redundancy*.

The images can be used in their digital form, stored and transmitted as files.

Depending on the size of image and on the necessary resolution, the amount of data generated by the digital images may be so great that results in impractical storage, processing and communication requirements. To reduce the image data, a solution would be to reduce its redundancy.

The term "data compression" refers to the process of reducing the amount of data required to represent a given quantity of information. The reduction of the amount of data required to represent a digital image is accomplished by the "image compression", currently recognized as as an "enabling technology".

Image compression techniques fall into two broad categories:

 - formation preserving or lossless

 - lossy.

In the first category the methods allow an image to be compressed and decompressed without losing information. The second category provides higher levels of data reduction but result in a less than perfect reproduction of the original image.

Lossy compression or irreversible compression methods/techniques are suitable for applications where minor loss of fidelity is acceptable to achieve a substantial reduction in the bit rate - number of bits conveyed or processed per unit of time.

An interesting method/technique for lossy compression is the Transform Coding that transmits coded information about the Fourier Transform of the original image.

At the reception it is obtained the original image by doing the Inverse Fourier Transform of the received Fourier Transform.

In mathematics the *Fourier Transform* FT of a function f (x), the F [f (x)] is called the frequency domain representation of f (x). It shows/describes the frequencies of the sine and cosine waves, which are present in the function f(x).

The Fourier Transform FT is named in the honour of the French mathematician and physicist **Jean-Baptiste Joseph Fourier** (1768 –1830), best known for initiating the investigation of the *Fourier series* and their applications to problems of heat transfer and vibrations. Fourier series were introduced by Joseph Fourier for the purpose of solving the heat equation in a metal plate.

He made important contributions to the study of trigonometric series, after preliminary investigations by Madhava, Nilakantha Somayaji, Jyesthadeva, Leonard Euler, Jean le Rond d'Alembert and Daniel Bernoulli. He applied this technique to find the solution of the heat equation, publishing his initial results in 1807 and 1811 years, and publishing his 'Theorie analytique de la chaleur' in 1822 year.

The Fourier series has many applications in electrical engineering, vibration analysis, acoustics, optics, signal processing, image processing etc.

The study of Fourier series is a branch of *Fourier analysis*.

In mathematics, a *Fourier series* is a sum that represents a periodic function or periodic signal as a sum of simple oscillating functions, sine waves $\sin(x)$ and cosines waves $\cos(x)$ or equivalent complex exponentials $e^{ix} = \cos(x) + i\sin(x)$ [Euler's formula].

$$f(x) = \sum_n [a_n \cos(nx) + b_n \sin(nx)] \qquad f(x) = \sum_n c_n e^{inx}$$

The Fourier Transform FT of f(x) is:

$$FT[f(x)] = f(x)\, e^{-inx} = \sum_n c_n e^{inx}\, e^{-inx} = \sum_n c_n$$

From FT back to f(x) by Inverse of Fourier Transform IFT:

$$I\{FT[f(x)]\} = FT[f(x)]\, e^{inx} = \sum_n c_n\, e^{inx} = f(x)$$

Many functions or signals are defined in the two-dimensional space or x-y plane as function f(x,y):

$$f(x,y) = \sum_m \sum_n c_n c_m e^{inx}\, e^{imy} = \sum_m \sum_n c_m c_n e^{i(nx+my)}$$

Then the two dimensional Fourier Transform FT of f(x,y) is:

$$FT[f(x,y)] = f(x,y)\, e^{-i(nx+my)} = \sum_m \sum_n c_m c_n$$

From FT back to f(x,y) by the two dimensional Inverse of Fourier Transform IFT:

$$I\{FT[f(x,y)]\} = FT[f(x,y)]\, e^{i(nx+my)} = \sum_m \sum_n c_m c_n\, e^{i(nx+my)} \qquad I\{FT[f(x,y)]\} = f(x,y)$$

In image processing it is used the two-dimensional *Discrete Fourier Transform* DFT, since a computer image is composed by pixels. DFT is a sampled Fourier Transform and therefore does not contain all frequencies forming an image, but only a chosen set of samples, large enough to fully describe the spatial domain image. Considering the image a square of area size MxM containing the discrete/sample pixels, and the number of frequencies equal to the number of pixels, the image in the spatial domain (x,y) and Fourier domain (u,v) are of the same size.

The Discrete Fourier Transform DFT of f(x,y) is:

$$F(u,v) = DFT[f(x,y)] \sim \sum_{y=0}^{M-1}\sum_{x=0}^{M-1} f(x,y)\, e^{-j(ux+vy)}$$

From DFT back to f(x,y) by the two-dimensional *Inverse of Discrete Fourier Transform* IDFT:

$$f(x,y) = I\{DFT[f(x,y)]\} = I[F(u,v)] \sim \sum_{v=0}^{M-1}\sum_{u=0}^{M-1} F(u,v)\, e^{j(ux+vy)}$$

In some applications, image data of the two dimensional DFT is compressed and transmitted.

At the receiver applying to the received data the two dimensional IDFT is obtained the compressed image. In order to quantify the reduction in data-representation size produced by an image compression algorithm, it is used the term "image data compression ratio c" between the original image and the compressed image. In order to quantify the quality of the compressed image, it is used the term "mean square error MSE" of the compressed image with respect to the original image.

The interactive program, which manipulates matrices as its fundamental data objects,

MATLAB or **MATRIX LABORATORY** calculates the Fourier transforms with a faster algorithm called *Fast Fourier Transform* FFT if the number of pixels is a power of 2.

The Fourier Transform produces a complex number output expressed by *magnitude/spectrum & phase.*
For the most real images, visualizing the Fourier Spectrum it is observed that the components of larger energy cluster in the low spatial frequency region and *the most of the energy distribution of the image is about the centre of the image in the transform space.*

Would be useful to extract those components of larger energy. By applying the IDFT only to those components of larger energy will result the slightly attenuated version of original image in the spatial domain, which is the compressed image.

The extraction of larger energy components is obtained by multiplying the image in the Fourier domain with a filter, which can follow the shape of the maximum energy distribution to achieve optimum compression. Such filters would have different shapes, see FIG 2: circle, square, stars

r   r

FIG 2

Closest filters are described by a hypocycloid/asteroid curve (asteroid means starlike in Greek).

For the image compression is used an asteroid zonal filter in the Fourier transform space.

The filter is a simple adaptive zonal filter with shape adjustment to maximize the energy stored in the transform coefficients.

The general Cartesian coordinates representation of an asteroid curve is given by the function:

$$u^e + v^e = r^e \qquad \text{for } 0 < e < 2$$

where u,v are the coordinates in the Fourier Transform space and r is the maximum half length.

The corresponding zonal filter H(u,v) is given by :

$$H(u,v) = 1 \text{ for } |u^e + v^e| \leq r^e$$

$$H(u,v) = 0 \text{ for } |u^e + v^e| > r^e$$

The representation of the asteroid curves for four values of e [e=2, 1, 0.75, 0.5] gives the shapes of circle, square (shifted $\pi/4$) and stars:

For:	Equation:	Shape:
e=2	$u^2 + v^2 = r^2$	circle
e=1	$u + v = r$	square
e=0.75	$u^{0.75} + v^{0.75} = r^{0.75}$	star1
e= 0.5	$u^{0.5} + v^{0.5} = r^{0.5}$	star2

19

To analyse the compression process it is used the *compression factor* c, which in the present topic it is defined as the ratio of the number of pixels of the original image to the number of pixels of the compressed image.

To analyse the quality of the compressed images in the present topic, it is used the normalised *mean square error* MSE of the compressed image with respect to the original image:

$$\text{MSE} = \sum_{j=0}^{M-1}\sum_{i=0}^{M-1} (f^*(i,j)-f(i,j))^2 \,/\, \sum_{j=0}^{M-1}\sum_{i=0}^{M-1} f^2(i,j)$$

i,j = sample coordinates in real space (square image MxM)

f(i,j) = sampled spatial intensities of the original image

f*(i,j) = sampled spatial intensities of the compressed image

IMPLEMENTATION

To realize the compression of an image/photograph file it is used the interactive program MATLAB.
The process of compression is done by MATLAB Program, following more steps:

- It is taken a digital image of size 128x128 pixels by scanning a photograph
- It is visualized the digital image - FIG 3

  The image has black and white horizontally & vertically directed features.
- It is obtained the Discrete Fourier Transform DFT of the respective digital image
- It is visualized DFT spectrum - FIG 4

  DFT image has also 128x128 pixels and its centre is the point (64,64).

It is observable that the significant energy distribution in the frequency domain is around image centre, aria that should be preserved.

- Are done the filters as 128x128 Hilbert Matrix Hj of ones 1s and zeroes 0s with ones 1s in the interior of figures which can follow the shape of the maximum energy distribution in the Fourier domain.

The figures are circle, square, star1 and star2 with the centre point (64,64) given by the asteroid equation:

$(x-64)^e + (y-64)^e = r^e$ for **e** values of 2, 1, 0.75, 0.5.

FIG 3

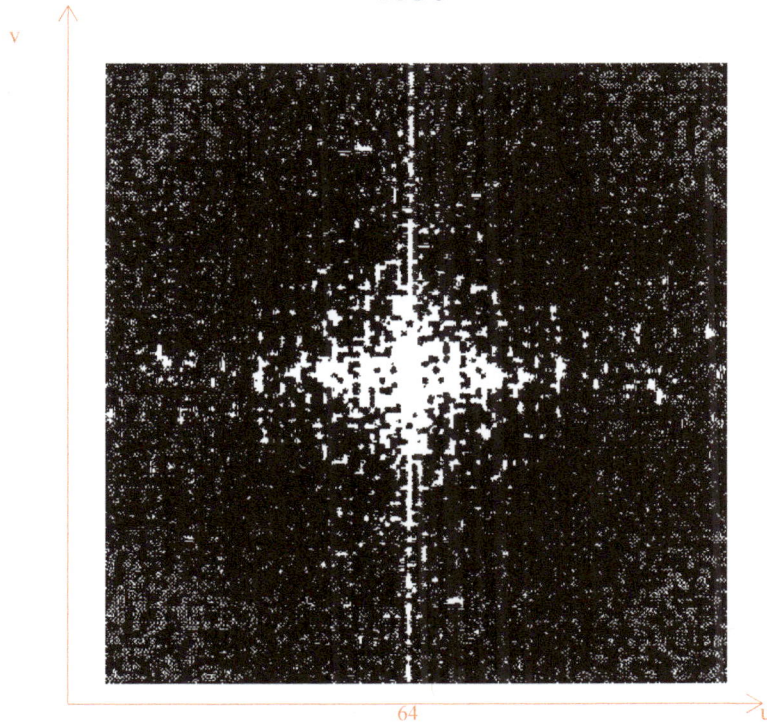

FIG 4

Sketches for Hilbert Matrix Hj shapes:

```
H1   [ 0 0 0 0 1 1 0 0 0 0        H2   [ 0 0 0 0 1 1 0 0 0 0
       0 0 1 1 1 1 1 1 0 0               0 0 0 1 1 1 1 0 0 0
       0 1 1 1 1 1 1 1 1 0               0 0 1 1 1 1 1 1 0 0
       1 1 1 1 1 1 1 1 1 1               0 1 1 1 1 1 1 1 1 0
       1 1 1 1 1 1 1 1 1 1               1 1 1 1 1 1 1 1 1 1
       1 1 1 1 1 1 1 1 1 1               1 1 1 1 1 1 1 1 1 1
       1 1 1 1 1 1 1 1 1 1               0 1 1 1 1 1 1 1 1 0
       0 1 1 1 1 1 1 1 1 0               0 0 1 1 1 1 1 1 0 0
       0 0 1 1 1 1 1 1 0 0               0 0 0 1 1 1 1 0 0 0
       0 0 0 0 1 1 0 0 0 0 ]             0 0 0 0 1 1 0 0 0 0 ]


H3   [ 0 0 0 0 1 1 0 0 0 0        H4   [ 0 0 0 0 1 0 0 0 0 0
       0 0 0 0 1 1 0 0 0 0               0 0 0 0 1 0 0 0 0 0
       0 0 0 1 1 1 1 0 0 0               0 0 0 1 1 1 0 0 0 0
       0 0 1 1 1 1 1 1 0 0               0 0 1 1 1 1 1 1 0 0 0
       1 1 1 1 1 1 1 1 1 1               1 1 1 1 1 1 1 1 1 1
       1 1 1 1 1 1 1 1 1 1               0 0 1 1 1 1 1 1 0 0
       0 0 1 1 1 1 1 1 0 0               0 0 0 1 1 1 0 0 0 0
       0 0 0 1 1 1 1 0 0 0               0 0 0 0 1 0 0 0 0 0
       0 0 0 0 1 1 0 0 0 0               0 0 0 0 1 0 0 0 0 0
       0 0 0 0 1 1 0 0 0 0 ]             0 0 0 0 1 0 0 0 0 0 ]
```

Considering constant either the *frequency bandwidth* or the *compression factor*
in the Fourier domain, there are studied two digital image compression cases.

Each case is studied for all filter shapes given by the four values of e: 2, 1, 0.75, 0.5

As a result there are two MATLAB programs: A & B

Program A is developed for *constant frequency bandwidth* in the Fourier domain: r=ct=64 Program A uses the equations:

$e=2 \qquad (x-64)^2+(y-64)^2=64^2$

$e=1 \qquad x+y=64 \quad x-y=64 \quad x-y=-64 \quad x+y=192$

$e=0.75 \quad (x-64)^{0.75}+(y-64)^{0.75}=64^{0.75}$

$e=0.5 \qquad (x-64)^{0.5}+(y-64)^{0.5}=64^{0.5}$

Program B is developed for *constant compression factor* between the original and compressed image in the Fourier domain: c = 128x128/Pj = ct =6

Program B uses the equations:

$e=2 \qquad (x-64)^2+(y-64)^2=29^2$

$e=1 \qquad x+y=92 \quad x-y=36 \quad x-y=-36 \quad x+y=165$

$e=0.75 \quad (x-64)^{0.75}+(y-64)^{0.75}=44^{0.75}$

$e=0.5 \qquad (x-64)^{0.5}+(y-64)^{0.5}=63^{0.5}$

Each program, A or B, follows the same steps.

PROGRAM A STEPS

- It is done the filtering by multiplying the Discrete Fourier Transform DFT with the Hilbert matrix filter of program A (4 variants)
- It is visualized the DFT spectrum after its filtering (compressing) – FIG 5
- It is visualized the compressed image by taking the Inverse Discrete Fourier Transform IDFT of the filtered DFT – FIG 6

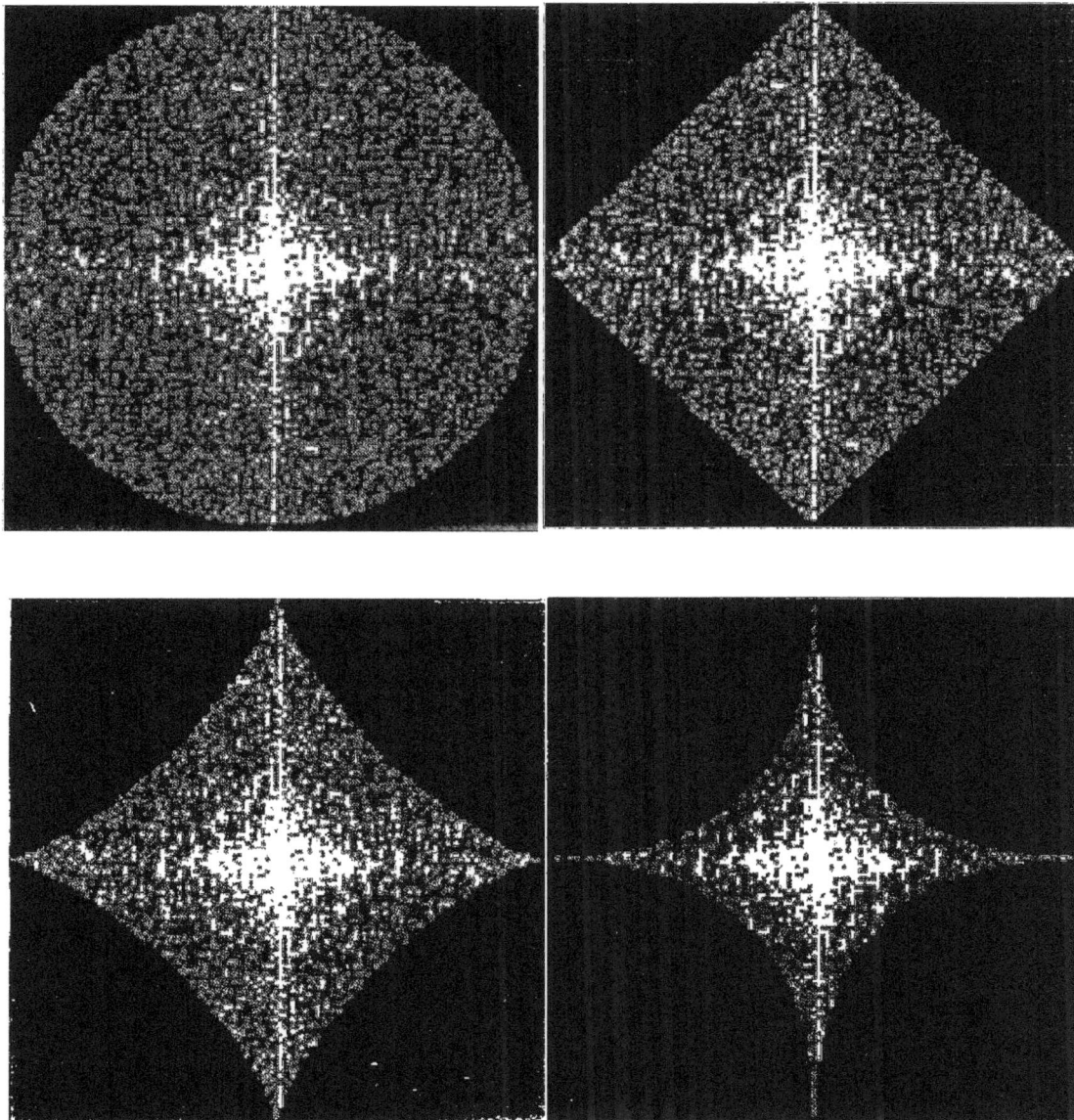

FIG 5

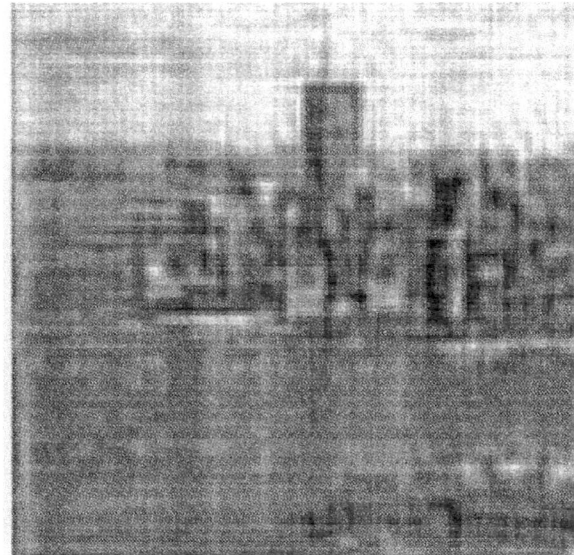

FIG 6

- It is calculated the realized compression factor c(e) and the error of filtering MSE(e) [r=64=ct.]

e	2	1	0.75	0.5
c	1.3	2	2.8	5.9
MSE	0.0027	0.0073	0.0107	0.0163

- It is done the graphic representation for c(e) and MSE(e) in FIG 7

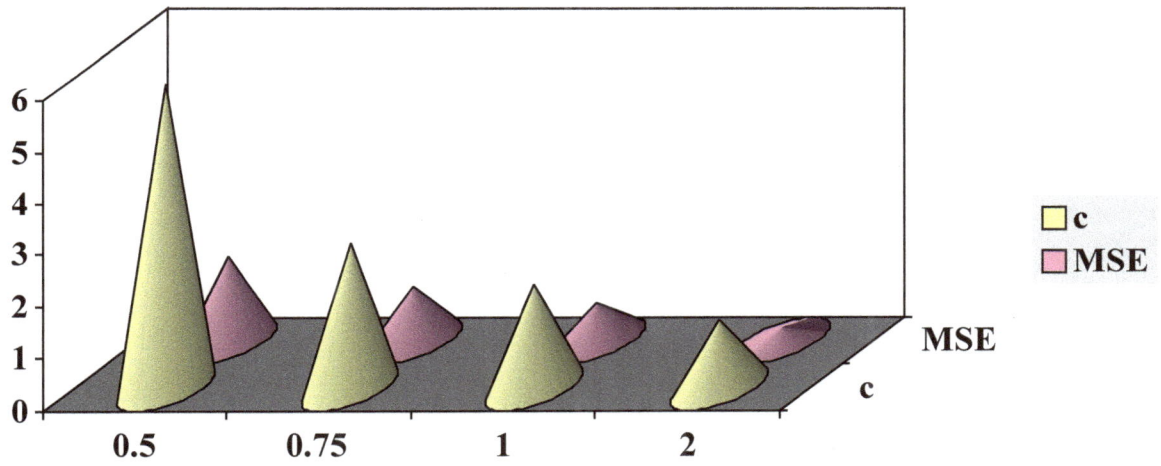

FIG 7

MATLAB PROGRAM FOR STUDY CASE A

```
path('h:\',PATH)
a=loadbmp('h:\newyork');
viewim(a)
ap=a(1:128,1:128);                    %square image
P=128*128                             %number of pixels of the digital image
viewim(ap)

b=fft2(ap);                           %Fourier transform of the image
bs=fftshift(b);
c=abs(bs);
m=max(c(:));
n=300*255/m;                          %normalization
c=n*c;
viewim(c)
bss=fftshift(bs);
a0=ifft2(bss);
a0=abs(a0);
viewim(a0)
D=sum(a0(:).^2)

for x=1:128                           %circular filter realized with a Hilbert Matrix
for y=1:128                           %[e=2  r=64]
H1(x,y)=1;
if y>sqrt(64.^2-(x-64).^2+64;
H1(x,y)=0;
end
if y>-sqrt(64.^2-(x-64).^2+64;
H1(x,y)=0;
end
end
end

d1=bs*H1;                             %filtering in frequency domain
d1a=n*abs(d1);
viewim(d1a)
d1s=fftshift(d1);
a1=ifft2(d1s);                        %compressed image
a1=abs(a1);
viewim(a1)

N1=sum((a0(:)-a1(:)).^2);
MSE1=N1/D                             %normalised mean square error
                                      %after filtering with the circular filter
P1=sum(H1(:));                        %number of pixels in filter
C1=P/P1                               %compression factor (circle-filter  r=64)
```

26

```matlab
for x=1:128                          %square filter realized with a Hilbert Matrix[e=1  r=64]
for y=1:128
H_2(x,y)=1;
if x<64;
if y<64-x;
H_2(x,y)=0;
end
end
if x>64;
if y<x-64;
H_2(x,y)=0;
end
end
if x<64;
if y>x+64;
H_2(x,y)=0;
end
end
if x>64;
if y>192-x;
H_2(x,y)=0;
end
end
end
end

d_2=bs*H_2;                          %filtering in frequency domain
d_2a=n*abs(d_2);
viewim(d_2a)
d_2s=fftshift(d_2);
a_2=ifft2(d_2s);                     %compressed image
a_2=abs(a_2);
viewim(a_2)

N_2=sum((a_0(:)-a_2(:)).^2);
MSE_2=N_2/D                          %normalised mean square error
                                     %after filtering with the square-filter
P_2=sum(H_2(:));                     %number of pixels in filter
C_2=P/P_2                            %compression factor (square-filter r=64)

for x=1:128                          %star filter realized with a Hilbert Matrix[e=0.75  r=64]
for y=1:128
H_3(x,y)=1;
if y<-(64.^0.75-(x-64).^0.75).^(4/3)+64;
H_3(x,y)=0;
end
if y>(64.^0.75-(x-64).^0.75).^(4/3)+64;
H_3(x,y)=0;
end
if y<-(64.^0.75-(64-x).^0.75).^(4/3)+64;
H_3(x,y)=0;
```

```
end
if y>(64.^0.75-(64-x).^0.75).^(4/3)+64;
H_3(x,y)=0;
end
end
end

d_3=bs*H_3;                              %filtering in frequency domain
d_3a=n*abs(d_3);
viewim(d_3a)
d_3s=fftshift(d_3);
a_3=ifft2(d_3s);                         %compressed image
a_3=abs(a_3);
viewim(a_3)

N_3=sum((a_0(:)-a_3(:)).^2);
MSE_3=N_3/D                              %normalised mean square error
                                         %after filtering with the star filter
P_3=sum(H_3(:));                         %number of pixels in filter
C_3=P/P_3                                %compression factor (star-filter e=0.75 r=64)

for x=1:128                              %star filter realized with a Hilbert Matrix[e=0.5 r=64]
for y=1:128
H_4(x,y)=1;
if y<-(64.^0.5-(x-64).^0.5).^2+64;
H_4(x,y)=0;
end
if y>(64.^0.5-(x-64).^0.5).^2+64;
H_4(x,y)=0;
end
if y<-(64.^0.5-(64-x).^0.5).^2+64;
H_4(x,y)=0;
end
if y>(64.^0.5-(64-x).^0.5).^2+64;
H_4(x,y)=0;
end
end
end

d_4=bs*H_4;                              %filtering in frequency domain
d_4a=n*abs(d_4);
viewim(d_4a)
d_4s=fftshift(d_4);
a_4=ifft2(d_4s);                         %compressed image
a_4=abs(a_4);
viewim(a_4)

N_4=sum((a_0(:)-a_4(:)).^2);
MSE_4=N_4/D                              %normalised mean square error
                                         %after filtering with the star-filter
P_4=sum(H_4(:));                         %number of pixels in filter
C_4=P/P_4                                %compression factor (star-filter e=0.5 r=64)
```

PROGRAM B STEPS

- It is done the filtering by multiplying the Discrete Fourier Transform DFT with the Hilbert matrix filter of program B (4 variants)
- It is visualized the DFT spectrum after its filtering (compressing) – FIG 8
- It is visualized the compressed image by taking the Inverse Discrete Fourier Transform IDFT of the filtered DFT – FIG 9

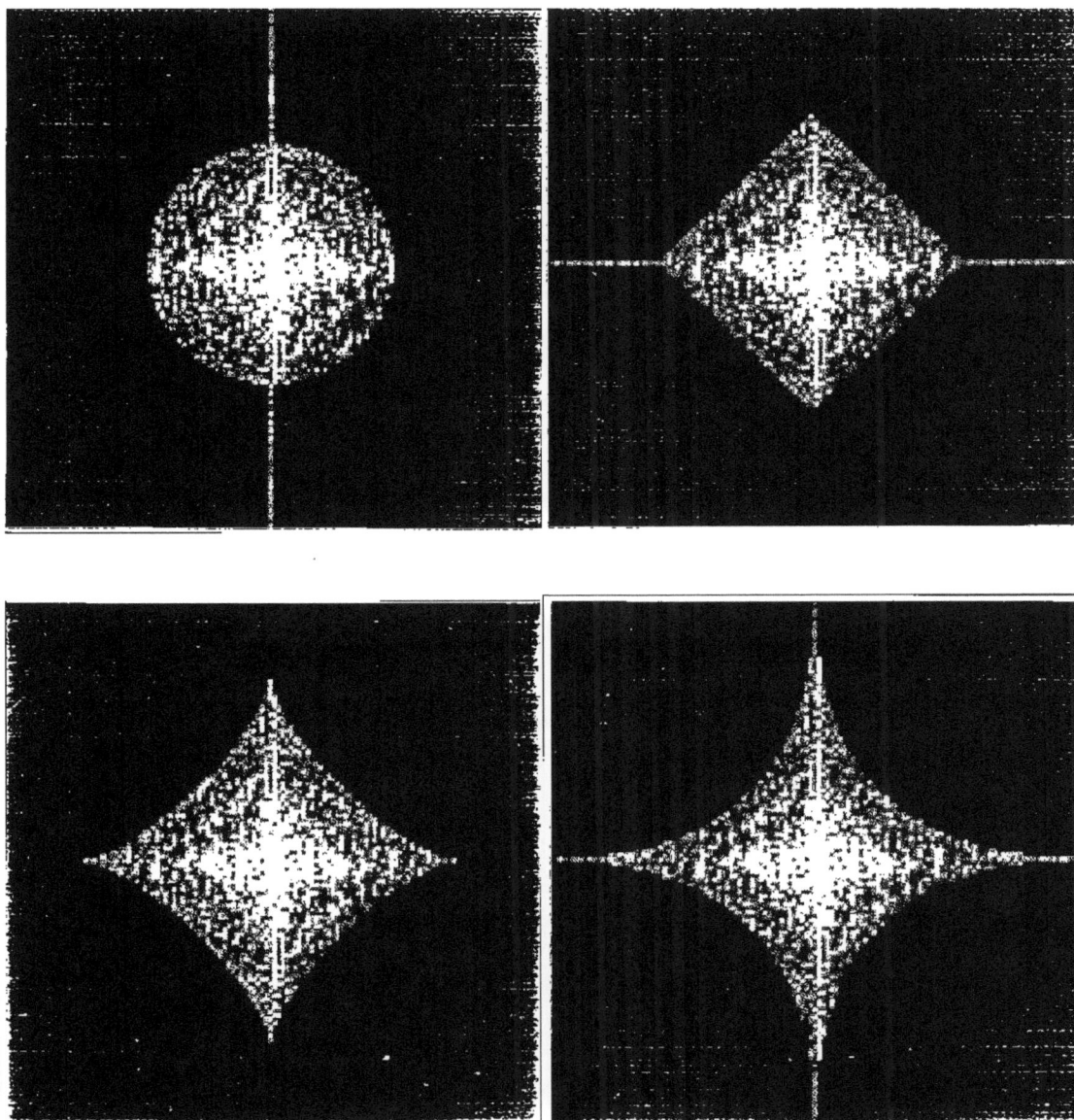

FIG 8

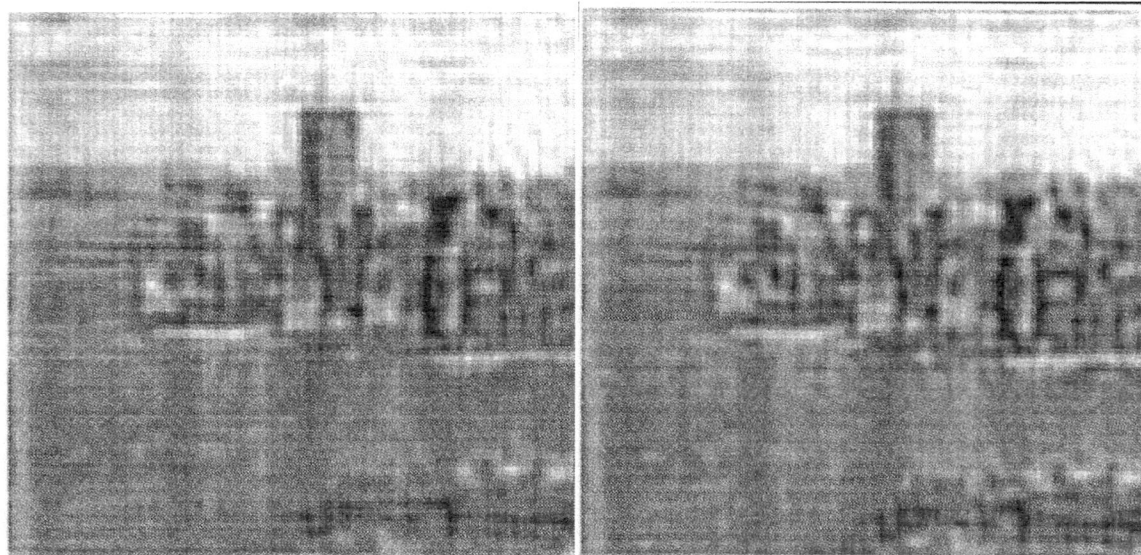

FIG 9

- It is calculated the error of filtering MSE(e) [c=6=ct.]

e	2	1	0.75	0.5
c	6	6	6	6
MSE	0.0177	0.0170	0.0168	0.0165

- It is done the graphic representation for c(e) and MSE(e) in FIG 10

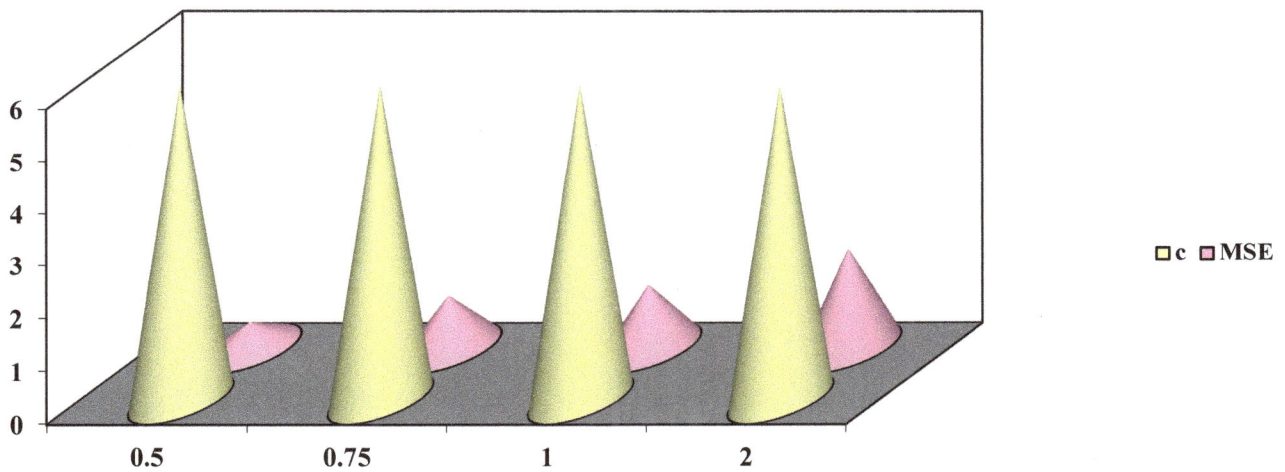

FIG 10

MATLAB PROGRAM FOR STUDY CASE B

```
path('h:\',PATH)
a=loadbmp('h:\newyork');
viewim(a)
ap=a(1:128,1:128);                    %square image
P=128*128                             %number of pixels of the digital image
viewim(ap)

b=fft2(ap);                           %Fourier transform of the image
bs=fftshift(b);
c=abs(bs);
m=max(c(:));
n=300*255/m;                          %normalization
c=n*c;
viewim(c)
bss=fftshift(bs);
a0=ifft2(bss);
a0=abs(a0);
viewim(a0)
D=sum(a0(:).^2)

for x=1:128                           %circular filter realized with a Hilbert Matrix[e=2  c=6]
for y=1:128
H1(x,y)=1;
if y>sqrt(29.^2-(x-64).^2+64;
H1(x,y)=0;
end
if y<-sqrt(29.^2-(x-64).^2+64;
H1(x,y)=0;
end
end
end

d1=bs*H1;                             %filtering in frequency domain
d1a=n*abs(d1);
viewim(d1a)
d1s=fftshift(d1);
a1=ifft2(d1s);                        %compressed image
a1=abs(a1);
viewim(a1)

N1=sum((a0(:)-a1(:)).^2);
MSE1=N1/D                             %normalised mean square error
                                      %after filtering with the circular filter
P1=sum(H1(:));                        %number of pixels in filter
C1=P/P1                               %compression factor (circle-filter  c=6)
```

```matlab
for x=1:128                              %square filter realized with a Hilbert Matrix[e=1  c=6]
for y=1:128
H2(x,y)=1;
if x<64;
if y<92-x;
H2(x,y)=0;
end
end
if x>64;
if y<x-36;
H2(x,y)=0;
end
end
if x<64;
if y>x+36;
H2(x,y)=0;
end
end
if x>64;
if y>165-x;
H2(x,y)=0;
end
end
end
end

d2=bs*H2;                                %filtering in frequency domain
d2a=n*abs(d2);
viewim(d2a)
d2s=fftshift(d2);
a2=ifft2(d2s);                           %compressed image
a2=abs(a2);
viewim(a2)

N2=sum((a0(:)-a2(:)).^2);
MSE2=N2/D                                %normalised mean square error
                                         %after filtering with the square-filter

P2=sum(H2(:));                           %number of pixels in filter
C2=P/P2                                  %compression factor (square-filter  c=6)

for x=1:128                              %star filter realized with a Hilbert Matrix[e=0.75  c=6]
for y=1:128
H3(x,y)=1;
if y<-(44.^0.75-(x-64).^0.75).(4/3)+64;
H3(x,y)=0;
end
if y>(44.^0.75-(x-64).^0.75).(4/3)+64;
H3(x,y)=0;
end
if y<-(44.^0.75-(64-x).^0.75).(4/3)+64;
H3(x,y)=0;
end
```

```
if y>(44.^0.75-(64-x).^0.75).(4/3)+64;
H3(x,y)=0;
end
end
end

d3=bs*H3;                               %filtering in frequency domain
d3a=n*abs(d3);
viewim(d3a)
d3s=fftshift(d3);
a3=ifft2(d3s);                          %compressed image
a3=abs(a3);
viewim(a3)

N3=sum((a0(:)-a3(:)).^2);
MSE3=N3/D                               %normalised mean square error
                                        %after filtering with the circular filter
P3=sum(H3(:));                          %number of pixels in filter
C3=P/P3                                 %compression factor (star-filter e=0.75 c=6)

for x=1:128                             %star filter realized with a Hilbert Matrix[e=0.5 c=6]
for y=1:128
H4(x,y)=1;
if y<-(63.^0.5-(x-64).^0.5).^2+64;
H4(x,y)=0;
end
if y>(63.^0.5-(x-64).^0.5).^2+64;
H4(x,y)=0;
end
if y<-(63.^0.5-(64-x).^0.5).^2+64;
H4(x,y)=0;
end
if y<-(63.^0.5-(64-x).^0.5).^2+64;
H4(x,y)=0;
end
end
end

d4=bs*H4;                               %filtering in frequency domain
d4a=n*abs(d4);
viewim(d4a)
d4s=fftshift(d4);
a4=ifft2(d4s);                          %compressed image
a4=abs(a4);
viewim(a4)

N4=sum((a0(:)-a4(:)).^2);
MSE4=N4/D                               %normalised mean square error
                                        %after filtering with the star-filter
P4=sum(H4(:));                          %number of pixels in filter
C4=P/P4                                 %compression factor (star-filter e=0.5 c=6)
```

CONCLUSIONS & COMMENTS

PROGRAM A - Constant frequency bandwidth of the image

The smallest normalised *Mean Square Error* MSE=0.0027 is obtained with circular filter for e=2. But for e=2 the compression factor is the smallest c=1.3.

The results show that the compressed images have better quality (smaller error) for circular filters but have also the smallest compression factor.

PROGRAM B - Constant compression ratio of the image

The smallest normalised *Mean Square Error* MSE=0.0165 is obtained with star filter for e=0.5.

The results show that the compressed images have better quality (smaller error) for star filters with smaller exponents e.

The edges and other sharp transitions in the grey levels of an image contribute significantly to the high-frequency content of its Fourier Transform.

All images showed in the presentation suffer after decompression of blurring or smoothing because of the reduction of high-frequency content.

For compressed images with low contrast it is necessary a further improvement, an enhancement, a contrast stretching.

references

TEORIA TRANSMISIUNII INFORMATIEI - ALEXANDRU SPATARU 1966

DIGITAL IMAGE PROCESSING - KENNETH R. CASTLEMAN 1979

AN ADAPTIVE ZONAL FILTER FOR DATA COMPRESSION - L.M.CHENG, A.S. HO, R.E.BURGE
1986

DIGITAL IMAGE COMPRESSION TECHNIQUES - MAJID RABBANI, PAUL W. JONES 1991

PRINCIPLES OF DIGITAL IMAGE SYNTHESIS - ANDREW S. GLASSNER 1995

DIGITAL IMAGING - MARK GALER, LES HORVAT 2002

THE COMPLETE GUIDE TO DIGITAL GRAPHIC DESIGN - BOB GORDON, MAGGIE GORDON
2002

DIGITAL IMAGE COMPRESSION TECHNIQUES - KASI V. GOMATHI, RAYAPPAN LOTUS 2014

A REVIEW ON IMAGE COMPRESSION TECHNIQUES - LALAANTIKA, ATISHSHARMA,
GAURAV GUPTA 2016

WIKIPEDIA

INTERACTIVE PROGRAM MATLAB

nD PRINTING

Printing as known from antiquity is a process for reproducing text and images, using a master form or template, by the means of printers. The things coming out of those printers are text and images that have been printed on a flat surface, flattish shapes or objects that cannot be picked up like two-dimensional (2D) artworks - drawings, paintings, photographs.

The process is called *2D Printing*.

The earliest examples of printers were *Seals* and *Stamps* used for making impressions.

The oldest seals come from Mesopotamia and Egypt. Back to early Mesopotamian civilization, before the year 3000 BC, the most common works of art to survive and feature complex and beautiful images were done by use of round *Cylinder Seals,* rolling an impress on clay tablets.

Brick Stamps for marking clay bricks, survive from Akkad from around the year 2270 BC.

There are also Roman lead pipe inscriptions of some length that were stamped.

And there is an unique gold foil sheet stamped with an amulet text from the 6th century BC.

Woodblock Printing, originating in antiquity in China, is a method for printing text, images or patterns that was used widely throughout East Asia. It was a way of printing on textiles and later on paper with surviving examples dating before AD 220.

The wood block, the printer, is carefully prepared as a relief pattern, which means the areas to show 'white' are cut away with a knife, chisel or sandpaper leaving the characters or image to show in 'black' at the original surface level. It is necessary only to ink the block by rolling over the surface with an ink-covered roller (brayer), leaving ink on the flat surface but not in the cut areas and bring it into firm, even contact with the paper or cloth to achieve an acceptable print. The content would print "in reverse" or mirror image, a complication when text was involved.

Woodblock Printing remained the most common East Asian method of printing books and other texts, as well as images, until the 19th century. In Japan the woodblock art print is called *Ukiyo-e*.

In Europe, the woodblock method for printing images on paper is covered by the term *Woodcut Printing* occasionally known as *Xylography.*

Intaglio Printing is another method of making prints invented in Germany by the 1430s years. Intaglio is the family of printing and printmaking techniques in which the image is incised into a surface and the incised line or sunken area holds the ink, direct opposite of a relief print.

As intaglio surface/matrix/printer were used the copper or zinc plates and the incisions were created by *Engraving*.

Block-book Printing is called the technique when both text and images are cut on a single wooden block for a whole page. A block book is a book printed from wooden blocks on which the text and illustration for each page had to be painstakingly cut by hand. The art of Block-book Printing is almost certainly of Chinese origin, probably of the 6th century AD.

The method had spread to Europe at least by the 15th century. Allan Henry Stevenson, an American bibliographer specializing in the study of handmade paper and watermarks, who "single-handedly created a new field: the bibliographical analysis of paper", by comparing the watermarks in the paper used in block-books with watermarks in dated documents, concluded that the "heyday" of block-books were the 1460s years. Block-books printed in the 1470s years were often of cheaper quality, as a cheaper alternative to books printed by "printing press".

Printing Press is a printer device for evenly printing ink onto a medium/substrate such as paper or cloth. The device applies pressure to the substrate that rests on its inked surface of *movable type* transferring the ink; *movable type* is the system of *printing & typography* using *movable pieces of metal type* allowing much more flexible processes than hand copying or block printing.

Typography is the art and technique of arranging type to make written language most appealing to learning and recognition, involving selecting typefaces, point size, line length, line-spacing (leading), letter-spacing (tracking) and adjusting the space within letters pairs (kerning).

Printing Press is a *movable printer*.

The world's first known movable type printing technology was invented and developed in China by the printer Han Chinese Bi Sheng between the years 1041-1048. Its invention and its spread are widely regarded as among the most influential events in the human history, revolutionizing the way people conceive and describe the world, guiding in the period of modernity.

In Korea, the movable metal type printing technique was invented in the early 13th century during the Goryeo Dynasty.

In Europe the invention of movable type mechanical printing technology is credited to the German printer Johannes Gutenberg in the year 1450. The high quality and relatively low price of the Gutenberg Bible printed in 1455 year established the superiority of movable type for western languages and printing presses rapidly spread across Europe, leading up to the Renaissance and later all around the world. Further along the years were developed also many other techniques for printing:

etching, drypoint, mezzotint, aquatint, lithography, chromolithography, rotary press, hectograph, offset printing, hot metal typesetting, mimeograph, photostat and rectigraph, screen printing, spirit duplicator.

Today, practically all movable type printing ultimately derives from Gutenberg's innovations for movable type printing which is often regarded as the most important invention of the second millennium. Modern printing is done typically with ink on paper but it is also frequently done on metals, plastics, cloth and composite materials. On paper it is often carried out as a large-scale industrial process and is an essential part of publishing and transaction printing.

Since the 1960s years, most types of high-volume books and magazines, especially when illustrated in colour, are printed with *offset lithography* (the inked oiled image is transferred or "offset" from a plate to a rubber blanket, then to the printing surface) that has become the most common form of printing technology.

The word *lithography* also denotes *photolithography*, a micro-fabrication technique used in the *microelectronics* industry to make integrated circuits and micro-electromechanical systems.

Computer Printing was done first time by the 19th century mechanically driven apparatus invented by the English mathematician, philosopher, inventor and mechanical engineer Charles Babbage for his *difference engine*, an automatic mechanical calculator designed to tabulate polynomial functions.

The most mathematical functions commonly used by engineers, scientists and navigators, including logarithmic and trigonometric functions, can be approximated by polynomials.

The automatic mechanical calculator used a series of metal rods with characters printed on them and a roll of paper stuck against the rods to print the characters.

An ancestor of the modern computer printer is the *stock ticker machine,* one of the first applications of transmitting text over a wire to a printing device based on the printing telegraph. Text typed on the typewriter at one end of the connection was displayed on the ticker machine at the opposite end of the connection. The "universal stock ticker", invented by the American inventor and businessman Thomas Edison in 1870 year, laid the basics for the *electric typewriter*. Pulses on the telegraph line made a letter wheel turn step by step until the correct symbol was reached and then printed at a very slow printing speed of one character per second.

Newer and more efficient tickers became available in the 1930s years, but these newer and better tickers still had an approximate 15-20 minutes delay. Ticker machines became obsolete in the 1960s years being replaced by computer networks.

A *ticker type electronic device* was produced in the year 1996 that could operate in true real time.

Digital Printing, developed in the second part of the 20th century, is referring to methods of 2D Printing on different materials from a file generated by the digital computer.
It has many advantages over the traditional methods.

Digital printing usually refers to professional printing where jobs from digital sources are printed on 2D, flat surface using different digital printers. Desktop publishing, variable data printing, fine art, print on demand, advertising, photos, architectural design are some of its applications.

Digital Printer is a peripheral in computing, a device used to take text and images from a computer and put them on paper engaging printing technologies as *blueprint, daisy wheel, dot-matrix, line printing, heat transfer, inkjet, electrophotography, laser, solid ink.*

The two most common printer mechanisms are black and white *laser printing* used for common documents and colour *inkjet printing,* used for high-quality output.

Laser Printing is an electrostatic digital printing process, first experimented in 1969 year, that very rapidly produces high-quality text, graphics and moderate-quality photographs.

Laser printers employ a xerographic printing process, alike used in *photocopy machines, multifunction / all-in-one inkjet printers* and *digital presses* which are slowly replacing many traditional offset presses in the printing industry for shorter runs.

The first commercial implementation of a laser printer was the IBM 3800, designed and manufactured by IBM company in the year 1976.

Inkjet Printing is a type of computer printing that recreates a digital image by propelling droplets of ink onto paper, plastic or other substrates. The concept of inkjet printing originated in the 19th century and the technology was extensively developed from 1951 year.

Inkjet printers are the most commonly used type of printer ranging from small inexpensive consumer models to very large expensive professional machines.

Starting in the late 1970s years, inkjet printers that could reproduce digital images generated by computers, were developed mainly by Epson, Hewlett-Packard (HP), Canon companies and later in year 1991 by Lexmark, spin-off from IBM company and by 2005 year, digital printing accounts for approximately 9% of the 45 trillion pages printed annually around the world.

Printed Electronics is the manufacturing of electronic devices using standard printing processes. It involves a set of 2D printing methods to create electrical devices on various substrates.

Electrically functional electronic or optical inks are deposited on the substrate creating active or passive devices as thin film transistors or resistors. Printed electronics technology can be produced on cheap materials like paper or flexible film, which makes it an extremely cost-effective method of production.

Printing typically uses common printing equipment suitable for defining patterns on material such as *screen-printing, flexography, gravure, offset lithography, inkjet.* Printed electronics is expected to facilitate widespread, very low-cost, low-performance electronics for applications such as *flexible displays, smart labels, decorative and animated posters, monitoring* and *active clothing* that do not require high performance.

Since early 2010 year several large companies are contributing heavily to the advancement of the printed electronics industry.

In a computer's workspace the actual dimensions are referred to the two-dimensional space 2D and the three-dimensional space 3D. 2D space is 'flat', using horizontal & vertical axes X&Y, the image has only two dimensions and if turned to the side becomes a line in one-dimensional space 1D. 3D space adds the third dimension Z, allowing for rotation and depth.

The difference between a representation in 2D space and a representation in 3D space is essentially like the difference between a painting and a sculpture: two-dimensional art is called painting and three-dimensional art is called sculpture.

In the year 1983, the American engineering physicist Charles Hull experimented with liquid acrylic-based materials, the photopolymers, which harden when exposed to ultraviolet UV light. *Stereo Lithography* was defined by him as a method for making solid three-dimensional 3D objects by successively "printing" thin layers of the ultraviolet sensitive material one on top of the other. Charles Hull built a machine that stacked layers of material to form a 3D object. The first object he printed in 3D, a cup about 5 cm tall, took months to produce. Charles Hull continued to make improvements to the machine and by the mid of 1980s years the machine was ready to print larger and more complex objects such as prototypes of machine parts.

3D Printing was born by the means of the newly created *3D Printer*.
The new 3D Printer attracted the attention of car manufacturers who wanted a way to create their own prototypes of parts such as door handles and stick-shift knobs.
Charles Hull patented his invention – U.S. Patent 4,575,330 entitled *Apparatus for Production of Three-Dimensional Objects by Stereo lithography* issued on March 11, 1986.
Further, Charles Hull founded the *3D Systems* company, in Valencia California in the 1986 year, whose first 3D Printer appearing on the market in 1988 year and was bought by airplane and medical equipment manufacturers and by several carmakers companies including General Motors or Mercedes Benz.
3D Systems company also developed the suitable software. Designers use this software to create *Standard Tessellation Language* STL files; STL is a file format Hull invented that converts   Computer Added Design CAD files into directions that printers can read to produce 3D objects.
In the early1990 years *Stratasys*, a 3D printing company in Rehovot Israel, developed a 3D printer that used a process called *Fused Deposition Modelling* FDM (plastic is melted and deposited in super-thin layer by layer through a heated nozzle) yielding durable models in comparison with the earliest method of Stereo lithography limited by the fragility of models and toxic chemicals. With FDM technology, the traditional fabrication process is substantially simplified, for example toolmaking becomes less expensive and time consuming, intricate designs that are impossible to make with conventional tooling are now possible and manufacturers realize immediate improvements in productivity,

efficiency and quality.

In the 2005 year the mechanical engineering professor at University of Bath, England,

Adrian Bowyer, founded *RepRap*, an open-source project to develop a 3D Printer that could print most of its own components. In February 2008 RepRap launched *Darwin 1.0*, a printer that could produce half of its components.

From 2012 year, some companies such as *Sculpteo* or *Shapeways* are proposing online solutions for 3D Printing.

3D Printing known also as *Additive Manufacturing* AM or *Desktop Manufacturing, Rapid Manufacturing, On-demand Manufacturing, Rapid Prototyping* refers to any of the processes for printing of three dimensional 3D objects using additive processes (in which successive layers of material are laid down) under computer control.

3D Printing applications now seem endless, beginning with jewellery and toys and continuing with robots, cars and even gourmet desserts that require precise layers of icing or chocolate. 3D Printing has also medical applications as custom prosthetics and scientists in Australia and the United States are now working toward printing artificial organs such as kidneys from patient's own cells.

Over the years were developed over 40 AM technologies.

New unique capabilities of AM technologies like shape complexity, hierarchical complexity, material complexity enable new opportunities for customization, very significant improvements in product performance, multi functionality and lower manufacturing costs; to take advantage of them, new design and CAD methods must be developed. As example of new capabilities are printed flat sheets of ceramic materials which can be rolled or folded into different shapes before firing and fuel cells in concentric circles or interwoven instead on the standard stack, creating more surface area and therefore easier transport of charge.

Objects of almost any shape or geometry can be produced by "3D Modelling" in a 3D Printer.

3D Modelling in 3D computer graphics is the process of developing a mathematical representation of any three-dimensional surface of an object via specialized software.

The product is called *3D Model*. It can be displayed as a two-dimensional image through a process called *3D Rendering* or used in a computer simulation of physical phenomena.

The model can also be physically created using 3D Printers.

The 3D Printer prints shaped layers that build up into a replica of the thing you want to copy or designed.

As materials, the industrial printers can use resin, clay, plastics or even powders of materials ranging from wood to metal. The year 2015 brings the first 3D printed car.

The real beauty of 3D Printing is that it removes constrains associated with traditional manufacturing, boosts project efficiency and reduces costs by allowing designs to be tested and adjusted quickly; it helps you think faster, innovate better and broaden horizons by providing

a new medium where creative minds can develop new applications for innovative engineering

in aerospace, architecture, automotive, commercial products, consumer products, defence, dental, education, entertainment, medical.

4D Printing is printing referring to the four-dimensional 4D space; 4D space adds the fourth dimension, T dimension, allowing for time.

On 27th February 2013, Stratasys company in collaboration with U.S. Massachusetts Institute of Technology MIT developed a 3D Printing with "smart" materials called the 4D Printing. Essentially, a 4D object is a 3D object evolving in time.

It is possible to create objects that have *four dimensions* 4D *characteristics* including a fourth dimension as a *dynamic component* that causes the structure of objects to change over time under the influence of water, heat, light.

This emerging technology will allow us to print objects that then reshape themselves or self-assemble over time: a printed cube that folds before your eyes, or a printed pipe able to sense the need to expand or contract. That means that those objects are made from *programmable materials.*

The American architect, artist and computer scientist Skylar Tibbits, as research scientist at the Massachusetts Institute of Technology, established what is known as the university's

Self-Assembly Lab, a cross-disciplinary research laboratory for inventing self-assembly and programmable material technologies aimed at reimagining construction, manufacturing, product assembly and performance.

Self-Assembly is a process by which disordered parts build an ordered structure through local interaction. The *concept of self-assembly on a nano-scale* has been around for years.

But now the *challenge is to demonstrate that this phenomenon is scale-independent* and can be utilized for self-constructing and manufacturing systems at nearly every scale and also to identify the key ingredients for self-assembly as a simple set of responsive building blocks, energy and interactions that can be designed within nearly every material and machining process available.

Self-assembly is a daring concept that would promise to enable breakthroughs across every application of biology, material science, software, robotics, manufacturing, transportation, infrastructure, construction, space exploration and the arts.

The courageous designers, scientists, engineers of the MIT Self-Assembly Laboratory have been researching for developing a variety of programmable materials as wood, textiles, flexible carbon fibre that have limitless applications.

5D Printing is built on 3D Printing, adding two more axis to the printing bed, beside the usual three XYZ axis, to allow a greater control over the printing process with the aim to create very complicated and detailed designs.

references

FORTUS 3D PRODUCTION SYSTEM - STRATASYS 2008

BE AMAZED 3D PRINTING-AN INDRODUCTION - STRATASYS 2014

LAYER-BY-LAYER: THE EVOLUTION OF 3-D PRINTING - AMANDA DAVIS 2014

3D PRINTED ROBOTS TEACH THEMSELVES TO MOVE - EVAN ACKERMAN 2014

ENGINEERS MUST INVENT INKS FOR MAKING 3D PRINTED FUEL CELLS -NEIL SAVAGE 2014

4-D PRINTING TURNS CARBON FIBRE, WOOD IN SHAPESHIFTING PROGRAMMABLE
MATERIALS - EVAN ACKERMAN 2014

PRECISION PROTOTYPING-THE ROLE OF 3D PRINTED MOLDS IN THE INJECTION MOLDING
INDUSTRY - LIOR ZONDER, NADAV SELL 2014

THE VALUE OF FDM FOR SAND CASTING - STRATASYS 2015

A NEW MINDSET IN PRODUCT DESIGN-3D PRINTING - STRATASYS 2015

4D PRINTING IS COOLER THAN 3D PRINTING AND WHY THAT MEANS THE END OF IKEA
FLAT PACKS - NICKY PHILLIPS 2015

ARTIFICIAL EVOLUTION, LEGGED MACHINES AND DELIVERY ROBOTS IN SILICON VALLEY
- EVAN ACKERMAN, ERICO GUIZZO 2016

THE BEST 3D PRINTERS OF 2017 - TONY HOFFMAN 2017

WHAT IS 5D PRINTING - TWI 2023

WIKIPEDIA

INTERNET of THINGS

INTERNET of EVERYTHING

Nowadays the information available on the Internet is, almost entirely, dependent on humans. The humans created the data found on the Internet, which is growing at high speed every second, by pressing a record button, taking a digital picture, scanning a bar code or by typing.

But a new concept has been in development for decades beginning in the early 1980s with the experiment in connection with a Coke machine at Carnegie Mellon University, a private research university in Pittsburgh, Pennsylvania, United States of America.

The experiment was performed by programmers, who could connect to the Coke machine over the Internet checking if there is a cold drink for them in case they decide to make the trip down to the machine.

The experiment led to the idea that equipping all objects in the world with minuscule electronic identifying devices or machine-readable identifiers would transform our daily life.

If all objects and people in everyday life would have tags or identifiers, they could be managed and inventoried by computers.

The *tagging* of things can be achieved electronically, using:

- radio-frequency identification
- near field communication
- barcodes
- quick response codes
- digital watermarking

Radio-Frequency Identification RFID is the wireless, non-contact use of radio-frequency electromagnetic EM fields to transfer data, for the purposes of automatically identifying and tracking tags attached to objects, not necessarily within the line of sight of the reader.

The tags containing electronically stored information use a local power source such as a battery or collect energy from the interrogating EM field and then emit microwaves or ultrahigh frequency UHF radio waves. Battery powered tags may operate at hundreds of meters.

An Electronic Product Code EPC is one common type of data stored in a tag.

When written into the tag by an RFID printer, the tag contains a 96-bit string of data:

- the first 8 bits are a header, which identifies the version of the protocol.

- the next 28 bits identify the organization that manages the data for this tag; the organization number is assigned by the EPC Global consortium.

- the next 24 bits are an object class, identifying the kind of product.

- the last 36 bits are a unique serial number for a particular tag.

The last two fields are set by the organization that issued the tag.

Like an URL (uniform resource locator/identifier or web address), the total electronic product code number can be used as a key into a global database to uniquely identify a particular product.

SHOP ENTRANCE

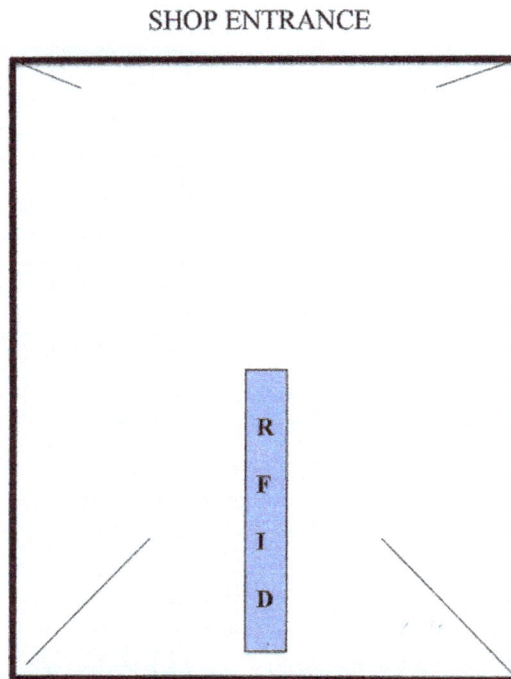

RFID at shop entrance

Near Field Communication NFC is a set of standards for smartphones and similar devices to establish radio communication with each other by touching them together or bringing them into close proximity, usually no more than a few inches. Communication is also possible between an NFC device and an unpowered NFC chip, called a "tag".

Present and anticipated applications include contactless transactions, data exchange and simplified setup of more complex communications such as Wi-Fi, a popular technology that allows an electronic device to exchange data or connect to the Internet wirelessly using radio waves.

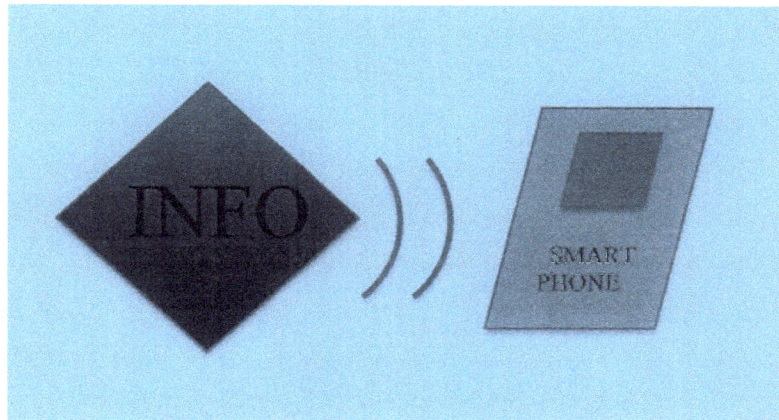

Transmission of information from an
Info-point to a mobile phone by NFC

Barcode is an optical, machine-readable representation of data in connection to the object to which it is attached, representing the data by varying the widths and spacing of parallel lines, one dimension patterns. Later they evolved into rectangles, dots, hexagons and other geometric patterns in two dimensions. Used in supermarket checkout systems the barcodes became very successful.

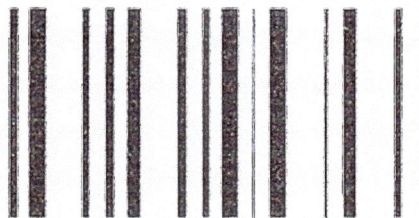

Barcode with parallel lines

The barcodes are scanned by barcode readers. Scanners and interpretive software are available on devices like desktop printers and smartphones.

Quick Response Code, first time designed for the automotive industry in Japan, is a type of matrix barcode or two-dimensional barcode that can contain up to 4000 characters, used to encode URLs, contact information, GPS coordinates or any free text.

A quick response code consists of black modules (square dots) arranged in a square grid on a white background, which can be read by an imaging device (such as a camera) and processed using Reed–Solomon error correction until the image can be appropriately interpreted; data is then extracted from patterns present in both horizontal and vertical components of the image. Applications include item identification, product tracking, time tracking, document management, general marketing. Originally designed for industrial uses, the quick response codes have become common in consumer advertising. Typically, a smartphone is used as a quick response code scanner, displaying the code and converting it to some useful form such as a standard URL for a website, avoiding to type it into a web browser.

Black modules used in quick response matrix barcode

Digital Watermarking is a kind of marker covertly embedded in a noise-tolerant signal. Also called *forensic watermark* is a sequence of characters or code embedded in a digital document, sound, image, video signal or computer program to uniquely identify its originator and authorized user. Forensic watermarks can be repeated at random locations within the content to make them difficult to detect and remove. Digital watermarks may be used to verify the authenticity or integrity of the carrier signal or to show the identity of its owners. It is prominently used for tracing copyright infringements and for banknote authentication. With a digital photo, a watermark is a faint logo or word(s) superimposed over the top of the photo.

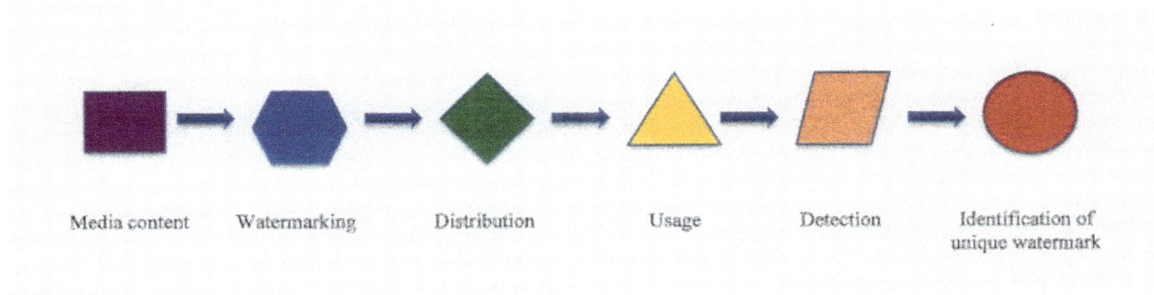

The watermarking process

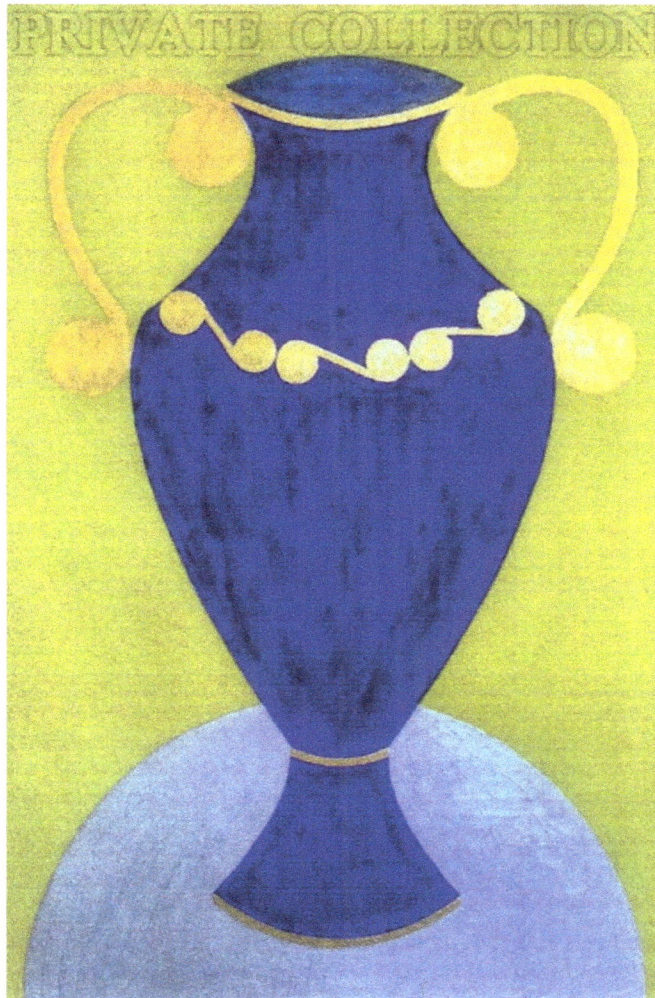

Image with upper watermark "PRIVATE COLLECTION"

Wireless Sensors are a very spread RFID companion technology.

Sensor is a converter that measures a physical quantity and converts it into a signal, which can be read by an observer or by an instrument. For example:

- a mercury-in-glass thermometer converts the temperature into the expansion and the contraction of a liquid, which can be read on a calibrated glass tube.

- a thermocouple converts the temperature to an output voltage (electrical force), which can be read by a voltmeter.

For accuracy, most sensors are calibrated against known standards.

At present many objects are becoming embedded with sensors and so they gain the ability to communicate. Sensors and actuators embedded in physical objects - from streets to pacemakers -are linked through wired and wireless networks, often using the same Internet Protocol IP that connects the Internet. So appear novel sensor networks and from there novel applications are going to play a starring role. Huge volumes of data flow to computers for analysis from the networks created by sensors.

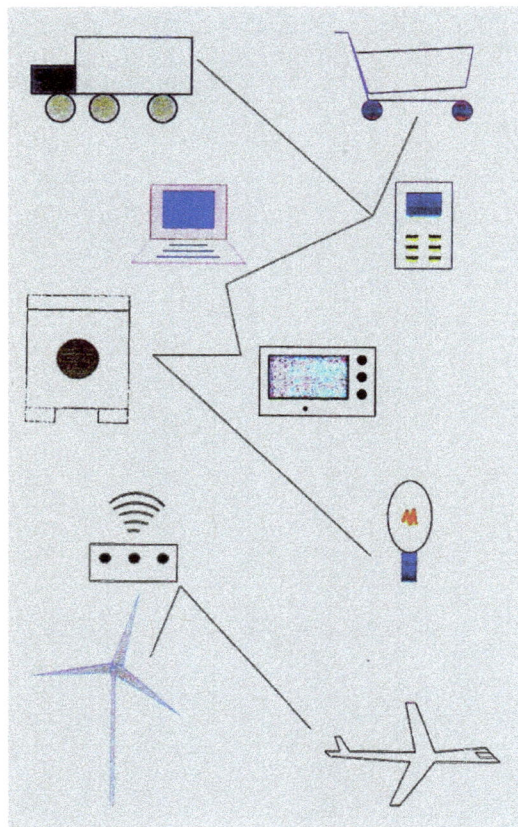

Objects communicating - M2M

When objects can both sense the environment and communicate, they become tools for understanding complexity and responding to it swiftly. We are now witnessing an explosion in these connected devices, both wired and wireless. Connected intelligent machines are a rapidly growing technology that embeds logic in devices to harness machine data and drive value for businesses. The result is an increase in the relevance of Machine-to-Machine M2M data,

opening up a huge arena of potential opportunities spanning diverse industries and applications. Products built with M2M communication capabilities are often referred to as being *smart*.

M2M refers to technologies that allow, both wireless and wired systems, to communicate with other devices of the same type and do not pinpoint specific networking or technology.

What's revolutionary in all the above is that these physical information systems are now beginning to be deployed and some of them even work largely without human intervention.

A scenario in which objects, animals or people are provided with unique identifiers and the ability to automatically transfer data over a network without requiring human-to-human or human-to-computer interaction is called **Internet of Things-IoT** *or* **Internet of Everything-IoE**.

The Internet of Things is a catch-all phrase used to describe the growing array of networked devices that are connected using the same protocols that power the Internet.

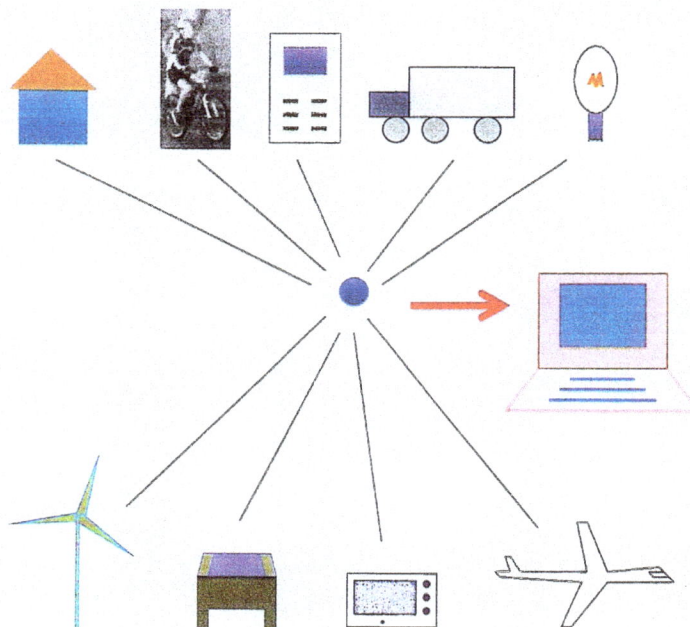

Things connected to Internet - IoT

*A **thing**, in the Internet of Things, or **everything**, in the Internet of Everything, can be a person with a heart monitor implant, a farm animal with a biochip transponder, an automobile that has built-in sensors to alert the driver when tire pressure is low or any other natural or man-made object that can be assigned an IP address and provided with the ability to transfer data over a network.*

IoT / IoE has evolved from the convergence of the *wireless technologies,* the *micro-electromechanical systems* MEMS and the *Internet.* It has been the most associated with machine-to-machine communication in manufacturing and power, oil and gas utilities.

In the next future this novel paradigm will play a leading role in logistics, intelligent transportation systems, business/process management, assisted living, e-health, to cite only a few examples.

On the Internet the method or protocol by which data is sent from one computer to another is the Internet Protocol IP.

Each host computer on the Internet has at least one IP address that uniquely identifies it from all other computers on the Internet.

The today most widely used version of the Internet Protocol IP, Internet Protocol Version 4 / IPv4 which uses a 32-bit address, allows approximately 4.3 billion addresses (2^{32}=4,294,967,296).

However Internet Protocol Version 6 / IPv6 is also beginning to be supported, providing much longer addresses and therefore the possibility for many more Internet users.

IPv6's huge increase in address space is an important factor in the development of the IoT.

IPv6 uses an 128-bit address, allowing approximately 3.4×10^{38} addresses,

$$(2^{128} = 340,282,366,920,938,463,463,374,607,431,768,211,456)$$

a huge number having in view that the number of atoms on surface of Earth is estimated to be 10^{36} atoms, the number of atoms contained in Earth is estimated to be 10^{50} atoms and the number of atoms in the entire observable universe is estimated to be within the range of 10^{78}-10^{82} atoms.

In other words, humans could easily assign an IP address to every *thing* or to *everything* on Earth.

Internet of Things is a world where physical objects are seamlessly integrated into the information network and can become active participants in business processes.

The connectivity to Internet in general is "wirelessly," but there's been a rapid evolution in existing technologies like cellular and WiFi, as well as an introduction of entirely new technologies dedicated to IoT. Wireless technology for IoT will be diverse.

Communication technologies would be: WiFi, Bluetooth, ZigBee, Z-wave, 6LowPAN, Thread, Celular, NFC, Sigfog, Neul, LoRaWan.

The term *Internet of Things* was proposed by the British technology pioneer Kevin Ashton in 1999 year. He is known for inventing the term to describe a system where the Internet is connected to the physical world via ubiquitous sensors.

The term is evolving as the technology and implementation of the ideas move forward.

The term Internet of Things generally refers to scenarios where network connectivity and computing capability extends to objects, sensors and everyday items not normally considered computers, allowing these devices to generate, exchange and consume data with minimal intervention. In attempting to define the Internet of Things we should keep in mind that it is fundamentally about communication, computation, sensing and actuation.

The emergence of the IoT has been described heralding a new wave of Internet connectivity and in that world, it is not longer about computers, it is not longer about people, it is about people and the world around them connecting.

Human beings in surveyed urban environments are each surrounded by 1000 to 5000 track-able objects. The Internet of objects would encode 50-100 trillion objects and be able to follow the movement of those objects.

Internet of things can be defined as a vision to connect everyday objects and devices to large databases and networks, using a simple, unobtrusive and cost-effective system of item identification and in the process, make them more intelligent and programmable.

IoT links uniquely identifiable things to their virtual representations in the Internet so IoT appears to move objects from the physical world to a virtual one.

By-products of this virtual continuum will be new services, functionality, applications.

Areas of applications include urban planning, sustainable urban environment, continuous care, emergency response, intelligent shopping, smart product management, smart meters, waste management, home automation and smart events.

The applications can be classified in six distinct types of two broad categories:

1. Information & analysis
    tracking behaviour
    enhanced situational awareness
    sensor-driven decision analytics
2. Automation & control
    process optimisation
    optimised resource consumption
    complex autonomous systems

The Internet of Things will be based on massive parallel IT systems (parallel computing) and the precise geographic location of a thing as its precise geographic dimensions will be critical.

Just as standards play a key role in the Internet and the Web, geospatial standards will play a key role in the Internet of Things.

The tremendous opportunities afforded by the IoT and driven by the rapid development in sensor technologies, like micro computing systems and wireless communication standards, mean that sensor device platforms will become a key element in enabling the connected world.

Some researchers argue that sensor networks are the most essential components of the IoT / IoE.

The IoT era brings new opportunities to traditional industries and drives business evolution for the next generation of products and services. To enable a diverse range of IoT applications, more efficient IoT development and to standardize different platforms and technologies, an open platform for IoT sensors and sensor nodes was established by sensor makers and module makers. Thai Advantech Corporation, along with British company ARM, German company Bosch Sensortec, Swiss company Sensirion and American company Texas Instruments, started a collaboration of a new Internet of Things (IoT) sensor platform called M2COM.

The Low Power Wi-Fi IoT Node WISE-1520 is the first module supporting M2COM.

The use of IoT technologies in manufacturing is called the *Industrial Internet of Things IIoT* or *Industrial Internet.*

By IIoT isolated programmable devices are transformed in networks of connected machines.

The driving philosophy behind the IIoT is that "smart machines" (intelligent devices using M2M technology like robots, self-driving cars and other cognitive computing systems that are able to make decisions and solve problems without human intervention) are better than humans at accurately, consistently capturing and communicating data. Instead to send data from these devices to a data center, the compute comes to those devices. That is called "fog computing".

Fog computing will bring intelligence and autonomy to industrial landscape.

The new term "smart city" has been emerging describing too broad numbers of technologies and applications. A *smart city* is an urban development vision to combine well information and communication technologies ICT and Internet of Things IoT solutions for managing a city's assets like schools, libraries, hospitals, transportation systems, power plants, water supply networks, waste management.

Busan is the second largest city in Republic of *Korea,* the 6th busiest container port in the world, with developed logistics, transportation and tourism. It is the first IoT-based *smart city* in Korea. It has well-equipped ICT infrastructure and strong IT technology. Using a cloud computing platform-as-a-service, the city connected the Busan Metropolitan Government, five local universities and the Busan Mobile Application Centre, which provides a setting for application development including workspace, offices, test equipment, APIs for public data and other tools. API is a software intermediary making possible for application programs to interact & share data.

South Korea also has built, at 65 Km SE of the capital Seul, the *smart city Songdo* from scratch. Nearly everything in this digital metropolis of smart homes is wired, connected and turned into a constant stream of data that is monitored and analysed by an array of computers with little or no human intervention. That proved that the Internet of Things or embedded intelligence in things with "smart systems that are able to take over complex human perceptive/cognitive functions, frequently acting unnoticeably in the background" can be a reality.

Amsterdam, capital of *Netherlands*, in 2013 year began to consider a huge opportunity, the benefits of a *smart city lighting system*, because the existing infrastructure, consuming 19% of
all electricity used, was decades old and a more efficient lighting could save $13.1 billion. Amsterdam had early initiatives geared toward public spaces, mobility and sustainability beginning in 2006 year.

Nice, the fifth most populous city of *France*, has developed a *smart city project* covering services as smart circulation, smart lightning, smart waste management and smart environmental monitoring. The goal of the smart city project in Nice was to "test and validate an IP-enabled technology architecture and economic model, as well as to determine the social benefits of IoE". The results can be used to help other cities accelerate smart city projects.

According to Juniper Research, Nice is the fourth smartest city in the world.

Nice hosts the annual TM Forum Live event, which includes smart city-related competitions.

TM Forum (TeleManagement Forum) is a non-profit industry association, for service providers and their suppliers in the telecommunications and entertainment industries.

The theme of the 2015 TM Forum "Hackathon" was *smart citizens in a smart city*.

"Hackathon" is an event typically lasting several days in which a large number of people meet to engage in collaborative computer programming. Occasionally, there is a hardware component as well. Groups compete with their results, which are sometimes selected and prizes are given. Internet of Parks, which uses IoT applications to service parks and municipal green spaces, won a Hackathon in year 2015, for its automated robot that cleans parks in Nice. The robots can pick up trash from trashcans, automatically order new trash bags and send an alert to operators if the device is experiencing a problem.

India in 2016 year, has launched the *government's smart city program*. It will implement

83 smart city initiatives in 20 cities, in an initial phase of the nationwide program.

The initial 83 projects require a total investment of INR 17.7 billion ($262.5 million).

According to ABI market research/intelligence firm based in New York, by 2020 year more than 30 billions devices will be wirelessly connected to the Internet of Things IoT or Internet of Everything IoE, the other term used.

According to Gartner Research, the demand for connected devices will go to one trillion by 2040.

The research further indicated that there will be US $166 billion invested in the IoT industry by 2020 year especially in the transportation, retail, warehousing, medical, and manufacturing sectors.

No matter what industry sector is as health, insurance, safety, government, or consumer related, the Internet of Things IoT has become their market's latest must-have.

American technology company Microsoft believes that the businesses can start with a few changes that make a big impact. It's about using your existing things in new ways and innovating and optimizing so everything works better together.

For this new, vast IoT ecosystem, Dutch international digital security company Gemalto proposes a solid *Foundation of Trust* around the three trusted pillars: trusted connectivity, trusted security, trusted monetization.

references

THE INTERNET OF THINGS - EDITORS: D. GIUSTO, A. IERA, G. MORABITO, L. ATZORI 2010 SMARTER

CONNECTED TECHNOLOGY - PANEL: T. BARR, J. DECAMP, A. HANSMANN 2013

DEMYSTIFYING THE INTERNET OF THINGS - JEFFREY VOAS 2016

3 SMART CITY USE CASES: AMSTERDAM, BUSAN AND NICE - SEAN KINNEY 2016

THE SMART GRID AND ITS ROLE IN INDUSTRIAL IOT - SEAN KINNEY 2016

IOT - JOHN GRIESING, AZIMUTH 2016

CREATE THE INTERNET OF YOUR THINGS - BOB KILBRIDE, MICROSOFT 2016

DRIVING THE IOT JOURNEY – ABI RESEARCH 2016

FOG COMPUTING IS THE FUTURE OF INDUSTRIAL IOT - STAN SCHNEIDER, RTI 2016

INDIA MOVES ON ITS SMART CITY PROGRAM - JUAN PEDRO TOMAS 2016

INTERNET OF THINGS IOT OUTLOOK FOR 2017 - MIKE KRELL 2017

WHAT IS THE INTERNET OF THINGS - BERNARD MARR 2017

GEMALTO, MINERVA, ORACLE, QUALCOMM, MCKINSEY&COMPANY RELEASES

LOGMEIN - MICHAEL SIMON

JOURNAL OF ADVANCED INTERNET OF THINGS

INSTITUTE FOR PERVASIVE COMPUTING - ETH ZURICH

THINGSQUARE - ADAM DUNKELS

WIKIPEDIA

SOLUTIONS OF
MODERN PROBLEMS

At present the most of the world countries strive to find solutions to solve chronic human society problems like homelessness, health-care inaccessibility, education inaccessibility, unemployment, tax evasion.

Their solving would create the opportunity of time and energy directed, by inventive minds, to new companies formation and the economy would flourish.

They can be solved:

1. *Tax System*

The companies should retain the tax from salaries and forward it to the Tax Office before they pay the workers.

The Tax Revenue will increase.

2. *Holidays*

The usual annual two weeks time holidays are not enough to eliminate the tiredness accumulated in the time of a year work and the tiredness is added year after year decreasing the work capacity.

The people involved in day after day work should care about their wellbeing by increasing the length of their holidays up to 3 months for example.

3. *Unemployment Solving*

The unemployed people should be on lists at companies in their profile, meaning depending on their qualifications or what they can do.

In the time of the employees' holiday, the companies should employ the people on the lists, people that should earn in that time more than an unemployment benefit for one year.

Due to this *holiday time employment* HTE the unemployment will disappear (see ANNEX).

If the number of jobs increases, the number of HTE jobs decreases and vice versa.

As example, at 6% unemployment rate would be 6.4% of jobs part of HTE.

The Tax Revenue will increase by not paying more unemployment benefits (hundred billion dollars per year).

4. *Health-Care System*

The entire health-care system, inclusive chiropractic and especially dentistry, using the latest achievements in science and technology in a humanitarian way, should be free of charge for everybody, poor or rich.

That would be financially possible because the Tax Revenue will increase (see point 3) and

that will allow allocating money for public health-care institutions, good protection against

the common enemy called "disease".

5. *Education System*

The entire education system, using the latest achievements in science and technology in a humanitarian way, should be free of charge for everybody, poor or rich.

That would be financially possible because the Tax Revenue will increase (see point 3) and

that will allow allocating money for public education institutions, good protection against

the common enemy called "ignorance".

6. *Living Opportunities*

With the increased Tax Revenue (see point 3) it is possible to invest in the building of houses

and flats, affordable for buying/renting to all people.

The spent funds will be reimbursed after the houses and flats are sold/leased.

That would be a challenging, humanitarian opportunity for global designers, architects,

engineers to show their talent, skill, ability, experience in new, modern, outstanding designs/projects, but also others can bring contribution.

The 21st century summarizes the rich knowledge and experience of the precedent centuries

and records an explosion of new information from a variety of human activity fields.

Nowadays, people of any profession can bring their priceless contribution on subject with opinions,

concepts, views, visions, solutions, sketches, designs, projects etc. in competitions, open to everybody,

organized to find the best, rewarding the winners with prizes.

Comments

As described above were found solutions for each chronic human society problem.
While the first solution about the tax evasion is independent from the others and the solution about unemployment
is also independent from the others, they are the only ones, which create money too. By far the most important is the
unemployment solution, which beside the fact that eliminates the unemployment, creates hundreds of billions of dollars yearly
(by not paying the unemployment benefits of hundreds of billions of dollars per year).
The health-care solution, the education solution and the homelessness solution consume money; they can be solved only with
money funds created mainly by the unemployment solution.
Briefly the solutions proposal has three important steps:

1. A new brochure should be sold for example in Post Offices, in what brochure all companies should advertise briefs about their
profile and kind of jobs they have (not only vacancies).
2. The unemployed people should appear on lists/apply at the companies in their profile, meaning depending on their
qualifications or what they can do.
3. A new government rule should appear in conformity with which the companies should employ the unemployed people on lists
in the desired work positions for 3 months. The existed occupants enter their 3 months holiday.
4. After all unemployed people are employed, the available country income money will increase by hundreds of billions of
dollars that will be directed to build free health-care, free education and affordable houses/flats.
The application of the solutions proposal will bring very important achievements:

- will eradicate the unemployment, creating for every citizen able to work a suitable workplace
- will increase the income and the moral of people because they will earn their own income and paid correctly in return higher
that the yearly unemployment benefit with no further obligations; also will increase the people income by the entire free of charge
health-care system, by the entire free of charge education system and by affordable living opportunities and even longer holidays
for rest/care.
For any necessary work should be employed people and paid correctly in return.
In general, voluntary work should be used in emergency situations only.
The solutions proposal is an invention and its application entitles the author to royalties.

ANNEX

Here is presented the calculation development of point 3 of presentation about the
holiday time employment HTE.
The unemployment rate U% *is a measure of the unemployment and it is calculated as a percentage*
by dividing the number of unemployed individuals by all individuals in the workforce.

The number of jobs affected by HTE would be H%.
H is given by formula: $H = 100\ U / (100 - U)$
Examples:

at 6.0 % unemployment rate	U= 6 H= 6.383	6.383% of jobs part of HTE
at 10.0 % unemployment rate	U=10 H= 11.111	11.111% of jobs part of HTE
at 20.0 % unemployment rate	U=20 H= 25.000	25.000% of jobs part of HTE
at 25.0 % unemployment rate	U=25 H= 33.333	33.333% of jobs part of HTE
at 33.33 % unemployment rate	U=33 H= 50.000	50.000% of jobs part of HTE
at 40.0 % unemployment rate	U=40 H= 66.666	66.666% of jobs part of HTE
at 50.0 % unemployment rate	U=50 H=100.000	100.000% of jobs part of HTE

The graphic in Fig 1 shows that for the unemployment rate 50 % (U = 50) all jobs become HTE jobs.

FIG 1

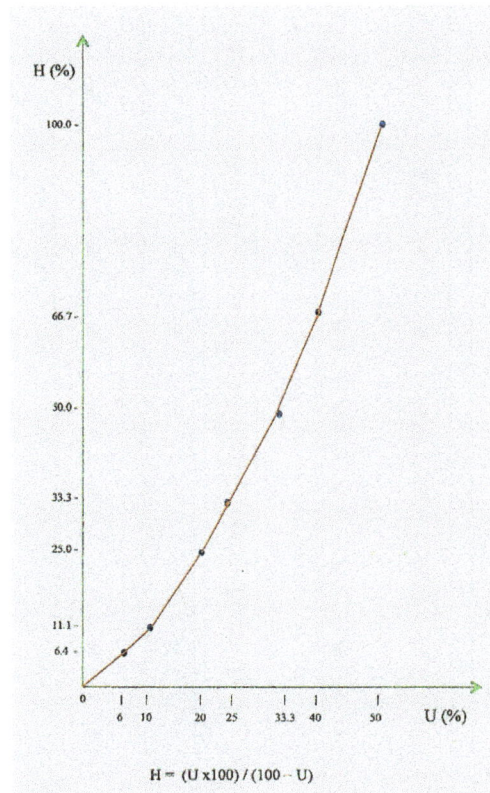

$H = (U \times 100) / (100 - U)$

If the unemployment rate is bigger than 50% (U > 50) would be necessary to develop the *double holiday time employment* DHTE.

The number of jobs affected by DHTE would be H'%.

H' is given by formula: $H' = 100\ U' / (100 - U)$ $U' = U - (100 - U) = 2U - 100$

Examples:

at 52 % unemployment rate	U=52	U'= 4	H'= 4.166	4.166% of jobs part of DHTE
at 55 % unemployment rate	U=55	U'=10	H'= 22.222	22.222% of jobs part of DHTE
at 60 % unemployment rate	U=60	U'=20	H'= 50.000	50.000% of jobs part of DHTE
at 65 % unemployment rate	U=65	U'=30	H'= 85.714	85.714% of jobs part of DHTE
at 66.66 % unemployment rate	U=66.66	U'=33.3	H'=100.000	100.000% of jobs part of DHTE

The examples show that for the unemployment rate 66.66% (U = 66.66) all jobs become DHTE jobs.

If the unemployment rate is bigger than 66.66 % (U > 66.66) would be necessary to develop the *triple holiday time employment* THTE.

The number of jobs affected by THTE would be H'' %.

H'' is given by formula: $H'' = 100\ U'' / (100 - U)$ $U'' = U - 2(100 - U) = 3U - 200$

Examples:

at 67 % unemployment rate	U=67	U''= 1	H''= 3.0303	3.030 % of jobs part of THTE
at 70 % unemployment rate	U=70	U''=10	H''= 33.333	33.333 % of jobs part of THTE
at 73 % unemployment rate	U=73	U''=19	H''= 70.370	70.370 % of jobs part of THTE
at 75 % unemployment rate	U=75	U''=25	H''= 100.000	100.000 % of jobs part of THTE

The examples show that for the unemployment rate 75 % (U = 75) all jobs become THTE jobs.

EARTH ENVIRONMENT
POLLUTION

Nowadays the main subject in media is the global warming, the warming of the entire earth, proved by 0.75 C degrees increase in the average temperature over the last century and by the largest rate of warming in the last 30 years.

The broad agreement among climate scientists is that the global temperatures will continue to increase leading to the global climate change of our planet Earth.

Although along the time the planet Earth encountered natural alternating periods of warming and cooling, many scientific, research, technological, climate, atmospheric, meteorological, astronomical, geological, geodesy, geophysics, physics, chemical, biology, microbiology, coral reef, oceanographic, foresters, wildlife, medical, health, statistical etc. societies agree that the human activities have become a major source of the latest environmental change.

The human activities are supposed to give to humans more than the nature can give them, respectively new materials, products, processes that would improve their life.

The humans invent and create taking examples from nature and are happy to obtain new things in different manufacturing processes, but do not care about what happen with the residual products and the used products that will accumulate as "debris" (the term "debris" depending on context).

Nature care about its own creations after they die, for example the autumn fallen leaves do not remain on soil stopping other plants to grow, because they are built from a material that quickly brakes down into its constituent chemicals which penetrate in soil nurturing the other plants.

But nature cannot decompose all materials created by humans.

That cannot be let at random/moods/voluntariate so a new project or design should not be considered ready until does also *design the procedure to remove the future "debris"*.

Any new project besides designing the main subject should contain the design of the way to remove the residual/undesired/used products of the manufacturing.

The design can involve automation and can create jobs.

Before the first industrial revolution (~ 1800 year) the producers used for the manufacturing of the new products the materials found in nature.

For the materials produced by it-self, nature has the ability to decompose them so the residual products of manufacturing and of course the used (not more needed or wasted) products are naturally decomposed.

As the sciences were emerging were tests/experiments to create and use new, man-made materials for new products but the good outcomes were in small quantities and the residual and wasted products were insignificant.

From the first industrial revolution the engines and the machines took the place of the horse/cattle power and the human force and the manufactured products increased very much in number and since were created new materials, not found in nature, with better proprieties from the human view point, the manufactured products increased very much in diversity too.

The social life was change in better as a consequence of the markets rich in new and useful products!

But the residual, undesired, used products resulted in the new context are not at all insignificant since they increased in quantity and in diversity too and not only that the natural decomposition last much more but can not occur at all at least at human life scale!

And so for over 100 years quantities of detritus, remnants, rubble, wreckage, garbage, junk, litter, refuse, waste are produced daily and are daily deposited on land, wasted in waters or spread in atmosphere and since their slow rate of decomposition they are doing in silence what is called the pollution of the environment.

By definition, to pollute the land, water, atmosphere means to make it dirty, impure and dangerous for people and animals to live in or to use, especially by means of poisonous chemicals that are produced as a waste product of an industrial process.

It is very interesting that in the context of the excitement provoked by the new, diverse and rich markets, the damaging effect of the wasted materials has passed undetected for more than one century!

Since the humans created the new materials they have to create also the means to remove them!

Doing a simple analysis of the present "debris", it can be divided into three categories, function of the state of aggregation: solid, liquid and gaseous.

– The solid "debris" usually is deposited on land or wasted in waters, where it can poison/stop the life of the plants and animals.

The solid debris appears to be the most conspicuous pollutant when visibly floats on the ocean's surface as "marine debris".

It should be removed from waters by contracted divers.

In the international waters the removing of "debris" should be done by schedule.

For the solid "debris" a means of getting rid of it is by *selecting* and *burning*.

And for example, the selected metallic parts (after melting) can be used to manufacture new metallic objects and the burning of the "debris" can be efficiently used in electric power stations.

– The liquid "debris" usually is wasted in waters, where it can poison and deplete them of plants and animals. It should be filtered and neutralized before enters the waters.

– The gaseous "debris" usually is spread in atmosphere, with effects like raising the temperature, changing the composition of the oceans and so on.

The heat from Earth is trapped into the atmosphere, due to the high levels of heat-trapping gases creating a phenomenon known as the "greenhouse effect".

In connection with the gaseous residual products is very important to mention the carbon dioxide CO_2, which is a direct result of burning fossil fuels (1.2 kg CO_2/1 kg coal), broad-scale deforestation and other human activities, because about half of the greenhouse effect is caused by CO_2.

Planting trees (man-made forests) remains one of the cheapest, most effective means of drawing excess CO_2 from the atmosphere, trees acting as a carbon dioxide sink by removing CO_2 from the atmosphere during photosynthesis (210 kg CO_2/ha/summer-day) to form carbohydrates that are used in plant structure/function and releasing oxygen back into air.

The industrial producers of CO_2 should be close to natural/man-made forests and should be realized a compromise between producing CO_2 by burning fuels and consuming CO_2 by forestation.

To care about the environment, to keep clean the planet Earth involve a restless work with wise decisions and flexible solutions, following the ways of nature rather than fighting them.

Since already the Earth environment is polluted, the administration of each village/town/city/region from every country should proceed to design the way to liberate their land, waters and atmosphere of the present pollutants, parallel with preventing the damaging effect of the new ones.

For a correct analysis of the present and future pollution, to find solutions and means to remove it, the administration should allocate financial funds and contract/employ temporary/periodically/permanently high level trained scientific, technological, research etc. staff.

The use of *electronics* based *devices to detect/measure/research pollution* is obvious.

A *"debris" processing industrial complex* could be a choice if/where necessary.

An international *establishment* concerned about *Earth Environment* based on the most advanced science and technology, with work done in background with humane, high level trained and paid staff, using the mankind knowledge accumulated in years of hard work and dedication, would assure a clean and reliable environment, the key to life of the present/future generations on Earth.

Reference: Wikipedia

November 2009

Since intense media appeals for opinions/solutions the writing was sent to
UNEP & UN Climate Change Conference 2009 / Copenhagen Summit 7-18 December 2009

Acknowledgement – EO 2508/2009

From: **Agnes Atsiaya** (agnes.atsiaya@unep.org) on behalf of **Executive Office**
          (executiveoffice@unep.org)
Sent: Tuesday, 17 November 2009 11:06:48 PM
To: irinarabeja@hotmail.com

Dear Ms. Rabeja,

On behalf of the Executive Director Mr. Achim Steiner, I wish to acknowledge and thank
you for your email dated 2 November 2009 regarding Environment Pollution and in the media.

Please note that your email has been brought to Mr. Steiner's attention.

Sincerely,

Paul Akiwumi
Chef de Cabinet
UNEP

UNITED NATIONS ENVIRONMENT PROGRAMME

Programme des Nations Unies pour l'environnement   Programa de las Naciones Unidas para el Medio Ambiente
Программа Организации Объединенных Наций по окружающей среде   برنامج الأمم المتحدة للبيئة
联合国环境规划署

UNEP

Our ref: 1159   12 May 2010

Dear Ms. Rabeja,

We thank you for your letter dated 1 December 2009.

This is to acknowledge and thank you for sending us information. Kindly note that
we keep the information in our file for future reference.

Sincerely,

Chizuru Aoki
Officer-in-Charge

Ms. Irina Rabeja
Electronics & computers consultancy
Email at irinarabeja@hotmail.com

cc: UNEP Executive Office (Susan Mutai and Jessica Wanyama)

Division of Technology, Industry and Economics
International Environmental Technology Centre (IETC)
2-110, Ryokuchi koen, Tsurumi-ku, Osaka 538-0036, Japan, Tel: +81.6.6915.4581; Fax: +81.6.6915.0304
1091 Oroshimo-cho, Kusatsu City, Shiga 525-0001, Japan, Tel: +81.77.568.4581; Fax: +81.77.568.4587
E-mail: ietc@unep.or.jp URL: http://www.unep.or.jp

STARLIGHT

FLEMING-MAURY-CANNON CLASSIFICATIONS
HERTZSPRUNG-RUSSELL DIAGRAM

INTRODUCTION

"The friendly gloss of stars on the night sky hides not only the Unknown but also the immense variety of the stars that often wander as double, triple and even quadruple, are either small as our Earth or big as our solar system, with an oscillating luminosity since they regularly expand and contract.
And although the life of the stars is measured in billions years, they age also and ultimately die or simply cool down until their gloss is weaker and goes off or contrary collapse and explode in a powerful fireball.
But still, while the old pass, new stars are born."

Our planet Earth is only a very small part of what we call the *universe*, commonly defined as the totality of everything that exists.

The word "universe" is derived from the Latin word "universum", which connects "uni" (one) with "versum" (rolled/rotated) used in the sense everything rolled/rotated as one or everything combined into one.

The universe is perceived by people from ancestral times, as the sky with stars visible at night by their spot. The starlight is the only thing that comes from or is given by universe to us.

But today from the starlight only the astronomers have knowledge about the attributes of the stars like: *chemical composition, surface temperature, age, absolute magnitude, luminosity, diameter, mass, volume, density, distance from Earth, velocity, speed of rotation, magnetic field.*

And that happened by observing and analysing only two unmistakeable/distinctive properties of the stars:

- *brightness* (study originated in antiquity by Hipparchus)

- *spectrum* (study done first time by Joseph von Fraunhofer)

They have opened the door to universe.

77

BRIGHTNESS OF STARS

Around the year 120 BC the Greek astronomers divided the stars visible to the naked eye into 6 classes function of their *brightness* B: from the brightest stars of class 1 to the faintest stars of class 6 at the limit of human visual perception. The brightness was measured in *magnitude(s)*, class 1 having magnitude 1, class 2 having magnitude 2… class 6 with magnitude 6. Each class had twice the brightness of the following class, a total range of 6 magnitudes in logarithmic scale.

Originated by the Greek astronomer, geographer and mathematician
Hipparchus (c.190 BC - c.120 BC), this method was popularized by the Greco-Egyptian mathematician, astronomer, astrologer, geographer and writer **Ptolemy** (AD c.100 - c.170) in his *Almagest*, a second century ancient Greek mathematical and astronomical treatise on the complex motions of the stars and planetary paths, its geocentric model accepted as dogma for more than
12 hundred years.

The English astronomer **Norman Robert Pogson** (1829-1891) formalized the system in
1856 year by defining a typical first class star (of magnitude 1) as a star that is 100 times brighter than a typical sixth class star (of magnitude 6), a magnitude 1 star being 2.512 times brighter than a magnitude 2 star, which is 2.512 times brighter than a magnitude 3 star and so on, where:
$$2.512 = \sqrt[5]{100} = \text{Pogson's Ratio}$$
The modern astronomy has kept in principle the same system, however is not limited to
6 magnitudes or only visible light, the scale has become enlarged (extended) to the extreme faint sky objects on one direction and on the other direction to the brightest heavenly bodies whose magnitudes are negative as Sirius (−1.4), full Moon (−12.74) or Sun (−26.74).
The Hubble Space Telescope has located stars with brightness of magnitude +30 to +31 at visible electromagnetic spectrum and Keck telescopes have located similarly faint stars in infrared electromagnetic spectrum.

The "electromagnetic spectrum" is the collective term for all possible frequencies of the "electromagnetic radiation / EM radiation / EMR", the radiant energy released by certain electromagnetic processes. Classically, the electromagnetic radiation consists of "electromagnetic waves", which are synchronized oscillations of the electric and magnetic fields, characterized by either the frequency or the wavelength of their oscillations, that propagate at the speed of light through a vacuum.

Visible light is one type of electromagnetic radiation; other types are invisible to the human eye, such radio waves, microwaves, infrared radiation, ultraviolet radiation, X-rays and gamma rays. The electromagnetic spectrum is the range of all types of EM radiation.

However the above presented magnitudes do not show the real light radiation of the star; they express a combination of the real spread of light of a star and its distance from Earth and so, a faint star would appear brighter at a shorter distance from Earth or the closest star to us Sun will appear a dot of light at enough farther distance. What is observed from Earth is the *apparent brightness* B_m measured in *apparent magnitude(s)* marked m, see FIG1. As measure for the real spread of light of a star, the astronomers defined the *absolute brightness* B_M measured in *absolute magnitude(s)* marked M, see FIG 2. That is the brightness of the same star situated at a standard distance of 10 *parsecs* from Earth, examples: *Sirius (1.4)* or *Sun (4.8)*.

The *parsec* meaning *parallax of one arc second* with the symbol *pc* is a unit of distance:
$1 pc = 31 \times 10^{12}$ Km $= 206,265$ AU ~ 3.26 light-years
One parsec corresponds to the distance at which the mean radius of the earth's orbit subtends an angle of one second of arc.

Knowing the distance D (in parsecs) between a star and Earth and the *apparent brightness* B_m of that star it is possible to calculate the *absolute brightness* B_M of that star with the formula:
$B_M = 5 + B_m - 5(\log D)$
To express scientifically the amount of electromagnetic energy radiated per unit of time by star (the power output of a star) it is used the term *luminosity* marked **L**.
Considering the luminosity of Sun as unit of measurement $L_{sun} = 1$, the luminosity of any star can be obtained by knowing its absolute brightness B_M with the formula:
$L = 10^{-0.4(B_M - 4.8)}$
Based on the **Stefan-Boltzmann Law**, stating that the radiant heat energy emitted from a surface is proportional to the fourth power of its absolute temperature, the luminosity of any star can be expressed as a function of its radius **R** and its surface temperature **T**:
$L = 4\pi R^2 \sigma T^4$
Higher the temperature and bigger the surface greater the energy flow in consequence the luminosity.

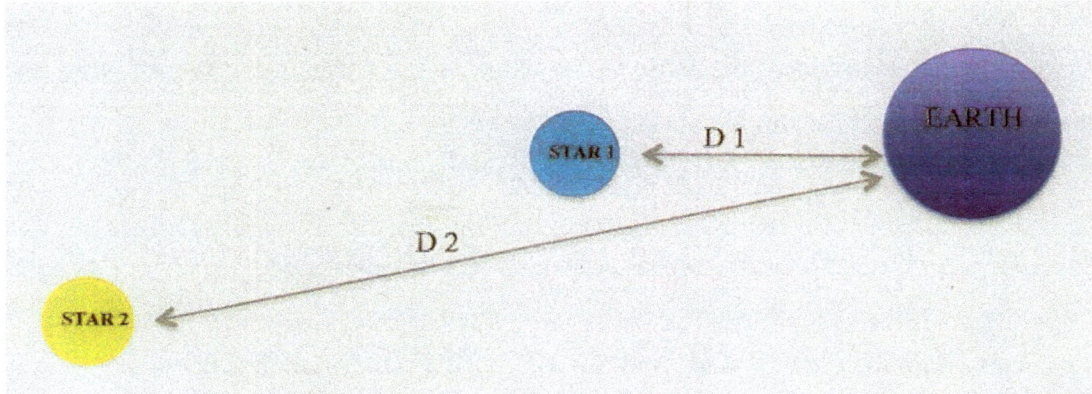

FIG 1 APPARENT BRIGHTNESS OF TWO STARS

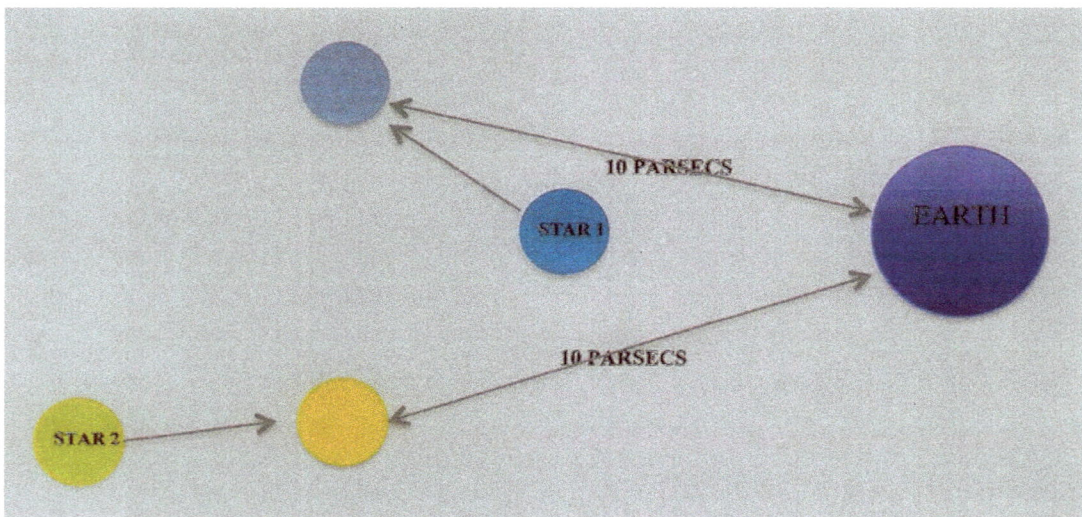

FIG 2 ABSOLUTE BRIGHTNESS OF TWO STARS

For two stars of luminosity L_1 and L_2 their luminosity ratio is function of their radius square ratio and the fourth power of their temperature ratio:

$$L_1 / L_2 = 4 \pi \sigma R^2_1 T^4_1 / 4 \pi \sigma R^2_2 T^4_2 \qquad\qquad L_1 / L_2 = R^2_1 T^4_1 / R^2_2 T^4_2$$

Our Sun has a luminosity of 3.84×10^{26} W (Js^{-1}) and a radius of 695 500 km.

In astronomical calculations it is often more convenient to consider SUN as unit of measurement for stars by stating:

unit of measurement for luminosity L_{sun}

unit of measurement for radius R_{sun}

unit of measurement for mass μ_{sun}

unit of measurement for temperature T_{sun}

Any other stars are compared with Sun.

The radius **R** of a star can be evaluated when it is in the previous relation with Sun by knowing its temperature **T** and its luminosity **L** and meaning $R_{sun}=1$, $L_{sun}=1$, $T_{sun}=1$:

$$R = (1/T^2) \sqrt{L}$$

Analysing the luminosity and the mass for different stars, was found that the luminosity of a star is bigger when its provision of energy is bigger, meaning its mass is bigger.

Resulted the relation between mass μ and luminosity **L**, mass-luminosity formula:

$$L = \mu^{3.5}$$

In consequence the *mass* μ of a star can be obtained knowing the luminosity **L** of that star:

$$\mu = \sqrt[3.5]{L}$$

So from the observed or apparent brightness of a star B_m it is possible to calculate for that star:

absolute brightness $\qquad B_M = 5 + B_m - 5(\log D)$

luminosity $\qquad L = 10^{-0.4((BM)-4.78)}$

radius $\qquad R = (1/T^2)\sqrt{L}$

mass $\qquad \mu = \sqrt[3.5]{L}$

SPECTRUM OF LIGHT

The light began to reveal its secrets in the year 1666 when the genial English mathematician and Nature scientist **Isaac Newton** (1643-1727) directed a ray from Sun through a prism and saw that on the wall of his room appeared the colours of the rainbow.

The prism decomposed the light in a row of colours entering lightly one in another from red, through orange, yellow, green, blue, indigo to violet.

To name the multicolour band that appeared like by magic on the wall of room, Newton took from Latin language the word *spectrum* meaning "host appearance" or "phantom" (plural *spectra* or *spectrums*).

Isaac Newton proved, what other thinkers agreed before, that the white light unifies in itself all the colours of the rainbow. The set of the 7 main colours in the light spectrum is given in FIG 3.

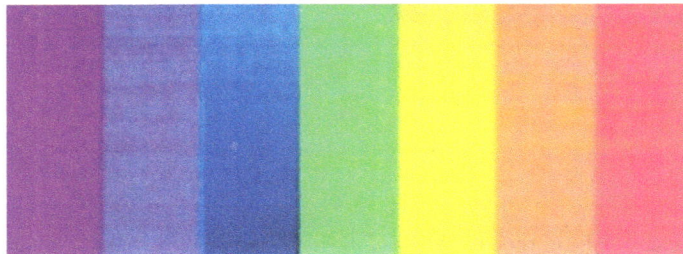

FIG 3   MAIN COLOURS IN LIGHT SPECTRUM

In 1900 years researchers studied in laboratory the spectrum of the light coming from the flame of different glowing burning gases, found bright lines of different colours and called them spectral lines of the emission spectrum, see FIG 4.

FIG 4   EMISSION SPECTRUM

The researchers also found dark lines in the colours of spectra of Sun and other stars and called them the spectral lines of the absorption spectrum, see FIG 5.

FIG 5   ABSORPTION SPECTRUM

The many dark lines in spectra of stars have a chemical origin; they are generated by absorbing elements and their observations and analyses led to the discovery of the code of *cosmic chemistry.*

The astronomers have developed a method to identify the chemical elements that generate the dark lines in spectrum by comparing carefully the spectrum of the starlight with the spectrum of different burning gases light obtained in laboratory.

In 1900 year the German physicist **Max Planck** (1858-1947) theorized that a hot body emits electromagnetic radiation in discrete quantities or quanta (later called photons).

A star is a hot body and the colours in the star spectrum are the visible part of the electromagnetic radiation of star with wavelengths from 380nm to 750nm, see FIG 6.

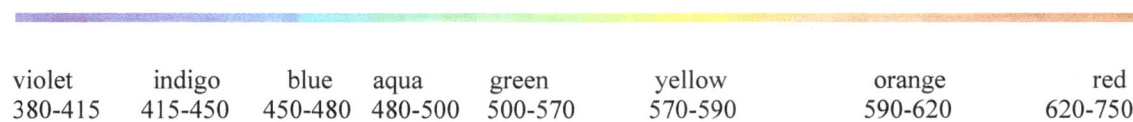

violet	indigo	blue	aqua	green	yellow	orange	red
380-415	415-450	450-480	480-500	500-570	570-590	590-620	620-750

FIG 6   THE VISIBLE SPECTRUM OR COLOURS OF STARLIGHT
AND THEIR WAVELENGTHS (nm)

A spectral line is a bright line or a dark line in an otherwise uniform and continuous spectrum, resulting from excess or deficiency of photons in a narrow frequency range, compared with the nearby frequencies.

The system of classification for stars based on the presence and strength of various types of absorption lines in their spectrum is called the *spectral type.*

The large-angle telescopes of the astronomical observatories equipped with special prisms can take hundreds of star-spectra simultaneously as photo images.

The spectra of stars from *star cluster Hyades* are shown in FIG 7.

The Hyades, also known as Mellote 25 / Collinder 50 / Caldwell 41, is the nearest open cluster to the Solar System (151 light years) consisting of a roughly spherical group of 300-400 stars of same place of origin, chemical content, age, motion through space.

From the observers on Earth, the Hyades cluster appears in the constellation Taurus.

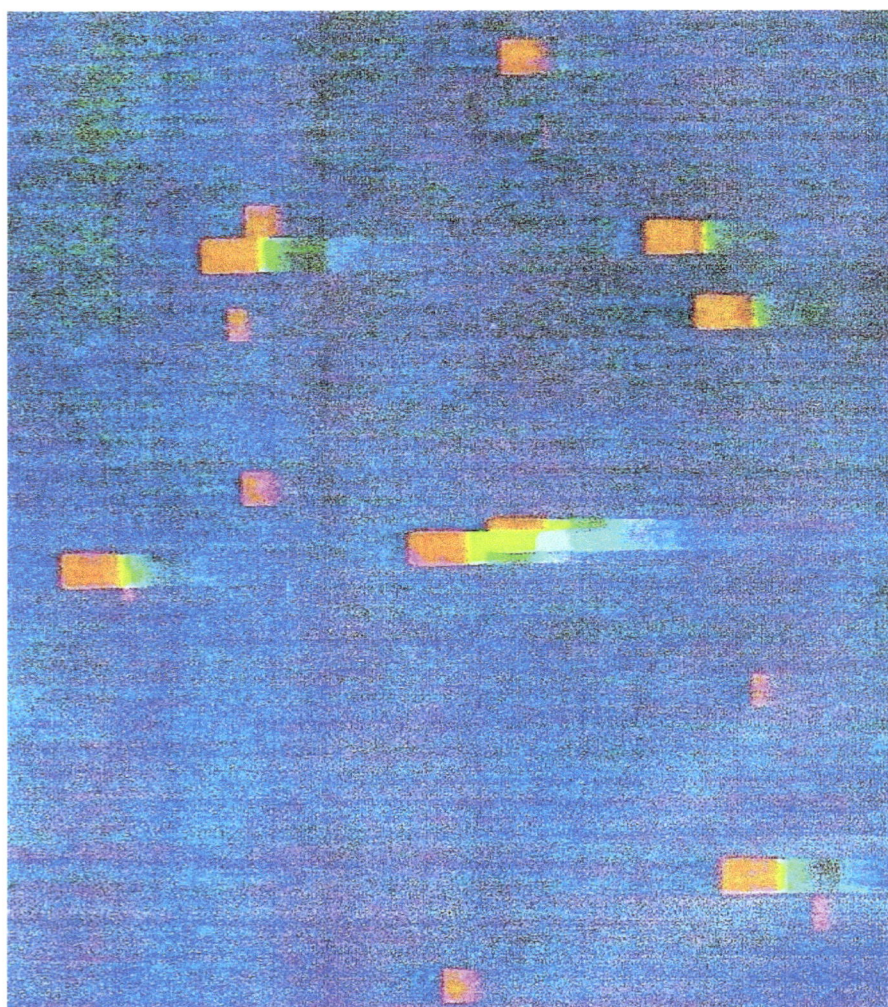

FIG 7   SPECTRA OF STARS FROM CLUSTER *HYADES*

The study of the spectrum of the light coming from a star gives us an abundance of information about that star like:

1.   The relative position of the spectral lines gives information about the *chemical composition* of the external gaseous layer of the star, because the spectral lines of a specific element or molecule always occur at same wavelengths and for this reason we are able to identify which element or molecule is causing the spectral lines.

2.   The chemical composition of a star indicates the star *age*. If a star contains a high percentage of elements other than hydrogen and helium, it is relatively young. Older stars contain few of these other elements. Our own star Sun is middle-aged compared to other stars.

3.   The *surface temperature* of a star can be estimated by studying its spectrum because the temperature determines the type and the number of present absorption spectral lines.
Few lines will indicate a hot star, and many lines will indicate a cool star.
The surface temperature of a star is also indicated by its colour.
Blue stars are generally hotter with temperatures around 40,000 K.
Red stars are cooler at less 3,500 K.
Our own star Sun which is yellow, has a surface temperature around 5,500 K.

4.   The intensity of certain spectral lines is a measure for the *absolute magnitude* of that star because the intensity of spectral lines show the abundance of respective elements in the light source, statistical correlations showing that the measured spectral line intensity is a function of the weight of material vaporised, more material more brilliant the star.

5.   The *luminosity*

6.   If the temperature and the luminosity of star are known can be simply calculated its *diameter,* the star diameters varying between few 100 million kilometres for the supergiants as *Betelgeuse* and few thousand kilometres for white dwarfs as *Sirius B*.

7.   From the luminosity it is possible to obtain (extract) the *mass* because they are in relationship, bigger the mass higher the luminosity. The masses of the stars do not vary too much, they are between tenth and ten times the mass of our own star Sun.

8.   Through the diameter of a star it is known its *volume* both having a big range of oscillation.

9.   The *density* of a star varies also correspondingly from hundred of thousandth of water density for giants to hundred of thousand times water density for white dwarfs and for neutron stars is even higher.

10.   The spectral determination of the luminosity of a star allows us to determine its *distance from Earth* since **Lambert Departure Law** says that there is a double distance to a light-cell if its brightness becomes half, so it is possible to calculate how far is a star by knowing the absolute brightness in comparison to the apparent brightness.

11.   The *velocity* at which the star is moving relative to Earth, can be determined by taking the **Doppler Effect** into consideration.
The Doppler Effect is a change in wavelength of the light, relative to the observer.
If the lines on a star's spectrum are shifted towards the red end of the spectrum, this means that the star is moving away from Earth. If the lines are shifted towards the blue end of the spectrum, the star is moving towards Earth.
The velocity is indicated by the amount that the lines shift: a large shift indicates a greater speed.

12.   The degree of extension of the spectral lines is a measure for the *speed of rotation* of a star around its axis. The rotation of a star causes the atoms on its surface to advance, retreat, or remain at a constant distance. This causes "smudges" on the spectrum, either towards the blue end or the red end. By measuring the width of these smudges, the scientists can determine the speed of rotation of a star.

13.   The split of the spectral lines shows a strong *magnetic field.*

BEGINNING OF MODERN ASTRONOMY

The German poet Johann Wolfgang von Goethe (1749-1832) said in the year 1790:
"The idea of white light being composed of coloured lights is quite inconceivable, mere twaddle, admirable for children in a go-cart".
The French philosopher Auguste Comte (1798-1857) was quoted as saying:
"There are some things of which the human race must remain forever in ignorance, for example the chemical constitution of the heavenly bodies".
What is cited above was the essence of a general opinion.
When in the 19th century the scientists deciphered and proved the code of light, namely that the light coming from stars gives us information about the stars composition, they opened the gate to universe.

Since then, for astronomers was accessible information contained in the light of stars and began an era like never before in science, when the scientists analysed, registered, classified hundreds of stars and on that was created the basis for progress to the theory of the stars realized in the 20th century.

Almost all what is known today about stars is based on *Spectroscopy*, the study of the interaction between matter and electromagnetic radiation, originating in the study of light dispersed by prism.

Isaac Newton (1642-1727) genial English physicist, mathematician, astronomer, natural philosopher, alchemist and theologian of the 17th century, placed himself near the earlier thinkers as the French **Rene Descartes** (1596-1650) proving that the white light unifies in itself all the colours of the rainbow.

William Wollaston (1766-1828) English chemist discovered in 1802 year that the Sun light does not form perfect spectrum, because dark lines slash the spectrum.

Josef von Fraunhofer (1789-1826) a young German optician in 1806 year seeking the best lens for telescope and other devices began the experiment of spectroscopy, the study of spectra, including the position and intensity of emission and absorption lines.
In 1814 year he made one of the earliest studies of the absorption lines. He installed a telescope in a dark room where he let the light fall through a narrow hole in the window blind, put a prism in the front of telescope and observed through the ocular that the resulting spectrum has a multitude

JOSEPH VON FRAUNHOFER

of vertical, strong and weak dark lines, some appearing even perfectly black. Those fine strips will be known in one day by any physics student as the *Fraunhofer Absorption Lines*.

Then he directed the new instrument named the *spectroscope* to the Moon, Venus and Mars.

The three spectra show the same order of lines like in the Sun spectrum and Fraunhofer deduced correctly that the heavenly bodies do not emanate their own light; they reflect the light of Sun.

When he directed his spectroscope to the Sirius and five other bright stars the spectral lines sample was for each star another and also different from the Sun sample, appearing that every star has its own spectrum sample, different each from other. He discovered that the spectra of various stars have different black lines. He hypothesized that the dark lines were caused by the absence of certain wavelengths of light.

Fraunhofer also invented a grate that gives a spectrum with much more detail than the prism.

After Fraunhofer, the scientists directed their search more on the bright than the dark spectral lines spending time more in laboratory than with the sky, work performed in many countries for three decades.

Gustav Kirchhoff (1824-1887) German physicist and **Robert Bunsen** (1811-1899) German chemist did a deciding experiment in the year 1859 at the University of Heidelberg trying to identify chemical substances by their colour in the time of burning and directing the light from the burner through a spectroscope to better differentiation.

So was proved that every chemical element when is burned as gas gives/shows an own typical sample/model of bright *lines of emission* in its spectrum.

Sodium, for example, delivers a pair of yellow lines that Fraunhofer observed in so many spectra because traces of this element are found in many substances.

By spectral analysis, Bunsen and Kirchhoff determined the sample of the coloured emission lines for all then known elements. Parallel they studied the flames in the Sunlight.

The *Fraunhofer Absorption Lines* of the Sunlight appear sharper and darker, obviously the gas taking more energy from the Sunlight than itself radiates.

Followed an astonishing conclusion: light from the hot Sun or other stars passes through their colder atmosphere; there the gases, like sodium steam, absorb their characteristic colour from light and so appear in the spectrum of the light arriving on Earth the dark *Fraunhofer Lines*.

On this basis they determined that in the Sun atmosphere also there are big quantities of iron, calcium, magnesium, nickel and chrome.

ROBERT BUNSEN

GUSTAV KIRCHHOFF

With the help of the new spectral analysis the astronomers began to collect/gather plenty of new knowledge about the stars.

William Huggins (1824-1910) a rich Englishman was one of the most zealous astronomers. He could publish in 1863 year with unique certainty:

Although the stars differentiate themselves by the way they look, however all are built in the same way as our Sun and contain matter, which at least partially coincides with the component elements of our solar system.

While men like Huggins revealed more and more star spectra other researchers developed more detailed systems to classify the spectral samples. Soon was shown that the stars can be classified in more classes or categories based on their *brightness, colour* and *spectral difference,* those classes helping to see the degree of similarity of the stars.

Angelo Secchi (1818-1878) Italian, Jesuit and physicist, Director of the Stars Department of Collegium Romanum in the course of five years examined the spectra of approximately 4000 stars overwhelming overlapped; however Secchi found enough similarity to classify them in four main spectral types function of their *colour, position, volume, number* and *darkness of the absorption lines* giving the *first useful, advantageous classification of stars* in 1868 year.

Henry Draper (1837-1882) a rich American, New York doctor and amateur astronomer was the first researcher to whom belong photographs of stars spectrums; already his father, also doctor and amateur astronomer, did the first photograph of Moon and one of the first photographs of the sun spectrum.

In 1872 year Henry photographed the spectrum of the star Wega and went on to record the spectra of over eighty other stars using the *spectrograph*, basically a spectroscope gifted with camera. Because of the sensitivity of the photo-plates towards the ultraviolet light the researchers could have pictures of the stars with so weak light that they are not observed with a telescope.

They could notice electromagnetic radiations at wavelengths not visible for the human eye.

In 1886 year Henry Draper's widow **Mary Anna Draper** set up a foundation in his honour; the aim of the foundation was to finance an ambitious programme for the spectrographic research and classification of the stars at Harvard College Observatory in Cambridge, Massachusettes for $400 000, eventually published as the **Henry Draper Catalogue**.

ANGELO SECCHI

HENRY DRAPER

Edward Charles Pickering (1846-1919) an American astronomer and physicist, the Director of Harvard College Observatory had the idea of that project.

Along with the German astrophysicist **Hermann Carl Vogel** (1841-1907) Pickering discovered the first spectroscopic binary stars and wrote *Elements of Physical Manipulations,* book in two volumes between 1873-1876 years.

Pickering invented and equipped the large-angle telescope of the observatory with special objective prisms taking hundreds of star-spectra simultaneously on 20x25 cm size photo-plates like the photo with spectra of stars from the star cluster *Hyades* presented in FIG 7.

And to bring in order the plates that have gathered in the headquarters of the observatory at the outskirts of Cambridge, he employed a group of women. After his opinion the women suit the best for tiring classifications and measurements.

WOMEN COMPUTERS AT HARVARD COLLEGE OBSERVATORY

EDWARD PICKERING

FLEMING - MAURY - CANNON CLASSIFICATIONS

Williamina Paton Stevens Fleming (1857-1911) immigrant Scottish teacher, Pickering's former housekeeper was his first leading assistant. Analysing photographic spectra which permitted fine divisions, the Williamina Fleming project together with Pickering developed schema extended and improved the old system of Secchi, which was based on the comparison of the samples of spectral lines, assigning to the stars twelve types marked with letters A - M but J, later completed with other types N O P Q R for new discovered stars or mixed spectra.

The first **Henry Draper Catalogue** was published in the year 1890, with 10,498 stars of the north half sky with a large fraction of work attributable to Flemming.

With the time, improved telescopes led to better wavelength resolution and some star types disappeared others merged. The last *Flemming's works published under her name* alone included her measurements of the apparent brightness of another 1400 stars as well as their spectral types.

Fleming examined all of the Harvard survey plates as soon as they were acquired, discovered 222 variable stars, 10 novae, discovered Horsehead Nebula and found in 1897 year the first spectrum of a meteor (published under Pickering's name).

Williamina Flemming was elected to honorary membership in the Royal Astronomical Society England.

Even before the first Draper Catalogue was published, Pickering had already identified other women to improve the Flemming-Pickering classification scheme for stellar spectra and put Antonia Maury to work on the stars of the Northern Hemisphere and Annie Jump Cannon to work on the stars of the Southern Hemisphere.

Antonia Coetana de Paiva Pereira Maury (1866-1952), niece of Henry Draper, held in 1888 year the mandate for the second classification project; she was probably the most gifted intellectually of the three women.

She studied at Vassar College, New York, USA with renowned professional astronomer **Maria Mitchell** (1818-1889) and graduated in 1887 year with honours in astronomy, physics and philosophy.

WILLIAMINA FLEMING

ANTONIA MAURY

Although she had been employed to classify objective-prism spectra into a system defined
by Pickering and Fleming, she instead set up her own system, which improved in two ways.

First was a finer gradation by the temperature of stars; Maury was first to recognise that the temperature
sequence must be $O\ B\ A$.

Second she noticed that in some cases the spectral features were unusually hazy (her type b)
and in some cases unusually sharp (her type c).

These and other details that Maury recorded were regarded by Pickering as a waste of time, attributing
them to the imperfection of instruments and it was not until about 1907 year that the Danish astronomer
Ejnar Hertzsprung who had independently discovered supergiants by another method, recognized the
importance of Maury's class c.

Many of Maury's b types were later recognized as rapid rotators, an interpretation she had
herself suggested.

In 1889 year Maury found the second spectroscopic binary star, b *Aurigae,* after Pickering discovered the
first, *Mizar,* the same year and she was the first to measure the orbital periods of both.

A binary star is one that consists of two stars gravitationally bound. Since their radial velocity
can be measured with a spectrometer by observing the Doppler shift of the stars' spectral lines, the
binaries detected in this manner are known as *spectroscopic binaries.*

Antonia Maury was a pioneer in the investigation of spectroscopic binaries.

Antonia Maury's own catalogue with the *a, b, c* characteristics and a variety of additional kinds of
information including notes of composite spectra and emission lines, appeared in 1897 year
in Harvard College Annals.

After 1892 year she was not longer formally employed by Harvard College Observatory but continued to
analyse spectra, her work appearing periodically in the Harvard Annals Bulletin
until 1935 year. Returned to Harvard College Observatory in year 1908 she published her most famous
work, the spectroscopic analysis of the binary star Beta Lyrae, in the 1933 year.

She was also a recognised ornithologist and naturalist.

Antonia Maury was the recipient of the Annie Jump Cannon Prize of the American Astronomical Society.

Annie Jump Cannon (1863-1941) trained in physics and astronomy at Wellesley College and Radcliffe College, Massachusetts, USA was Pickering's next assistant.

Appointed to the observatory staff in 1896 year, she spent her entire career there.

One of the most extensive efforts to classify the stars was Pickering's **Henry Draper Catalogue,** which provides the positions, brightness and spectra of 225,300 stars. An invaluable reference for astronomers covers the heavens from pole to pole for all stars brighter than the eighth magnitude, as well as many fainter stars and provides data on distances, distributions and motions.

Scientists investigating the colours, temperatures, sizes and compositions of stars frequently refer to the Henry Draper Catalogue for its spectral information.

The development of the catalogue was a colossal challenge with nearly a quarter of million stars to be classified. Pickering chose Cannon as principal investigator for the project.

In that capacity, she not only identified, recorded and indexed the data on the stars but also supervised the publication of all nine volumes in the year 1918.

Cannon personally examined every single one spectrum. Cannon revised the symbols used for the spectral types and reordered the classes in a more specific and subtle terms of decreasing surface temperature classifying the stars function of their *colour* - from bluest to red.

The main classes of the spectrum had only the sequence O B F G A K M and the letters of the sequence had a decimal division that gave to system a new precision. Sun, for ex. has the spectral class G2.

She devised the Draper classification scheme, which was introduced in her

Cannon's own Catalogue of the Spectra of 1122 Stars and was adopted internationally.

The Henry Draper catalogue together with Annie Cannon classification system became

standard works in astronomy.

Cannon's contribution in the field of spectroscopy was unsurpassed in quantity. Probably no other single observer in the history of science gathered so a great mass of data on a single system.

Cannon examined photographs of the stars near the South Celestial Pole for years and throughout her career she classified one-third of million stars and discovered more than 300 variable stars,

5 novae and many stars with peculiar spectra.

Cannon believed that patience not genius was responsible for her success.

Cannon won many honors for her work:

- Henry Draper Medal - for notable investigations in astronomical physics
- William Cranch Bond Astronomer - for her distinguished service at Harvard College Obs
- Ellen Richards Research Prize
- Doctor in Science degree - Oxford University - first woman in 600 years history

ANNIE JUMP CANNON

- Doctor in Law - Wellesley College
- Honorary degree - University of Groningen,Netherlands
- Honorary degree - University of Delaware
- Honorary degree - Oglethorpe University
- Honorary degree - Mount Holyoke College
- Honorary Member - Royal Astronomical Society England
- A Moon crater was named in her honor.

Cannon endowed the **Annie Jump Cannon Prize** of the American Astronomical Society.

**The wonderful work of Williamina Fleming, Antonia Maury and Annie Jump Cannon
in describing and ordering hundreds of thousands stars built the backbone of the modern
astronomy.**
Only by their work were found connections / relations between different features of stars.

Those connections / relations were developed and revealed short after the turn of the 20th century,
in a genial diagram by two astronomers from both sides of the Atlantic Ocean, independent one from
another, the Danish Ejnar Hertzsprung and the American Henry Norris Russell.
It was the HERTZSPRUNG - RUSSELL DIAGRAM.
The Hertzsprung - Russell Diagram owes its creation / apparition, first of all to the underestimated work
of Antonia Maury. To her was conspicuous that the relative width and darkness of some lines in spectrum
come from stars of same colour and spectral class and for that she built the subgroups *a, b, c*. Those
subgroups, especially the *c* subgroup waked the interest of the Danish astronomer Ejnar Hertzsprung.

HERTZSPRUNG - RUSSELL DIAGRAM

Ejnar Hertzsprung (1873-1967) was a Danish chemical engineer, who from the chemical process of photography came to star photography and from that to astronomy.

From 1909 year he worked at the observatories of Gottingen, Potsdam, Leiden.

In 1905 and 1907 years he published two papers both entitled "Zur Stralung der Sterne" (On Radiation of Stars) in a journal for scientific photography, in which he had used stellar colours determined from his own work and distances estimated from proper motions to show that stellar brightness came from two stars groups, which he called "Riesen" (Giants) and "Zwerge" (Dwarfs).

He concluded from his observations that the *luminosity* of a star is in relation with the *spectrum* and the *temperature* of that star, namely in the way that cold red stars in the same time have the lowest brightness. Hertzsprung recognition of very bright supergiants in 1905 year served to validate the spectroscopic criterion "c trait" identified by Antonia Maury and denied by Edward Pickering.

In 1908 year he sent a letter to Harvard College Observatory informing Pickering about the new subgroups and hurried him to consider his new subgroups in the Henry Draper classification but Pickering remained unimpressed.

In 1911 year the Danish astronomer Ejnar Hertzsprung plotted the absolute magnitude of stars against their colour in what became known as colour-magnitude diagram for the Hyades star cluster; the most of the stars arranged themselves in a narrow band marked by Hertzsprung as *Main Sequence* with the bright, blue stars left and the dark, reddish stars right; the exception were the red giants which built an own group.

Hertzsprung carry out astronomical work until about three years before his death, including determination of brightness of stars and their colours by photography, discovery of the variability of Polaris, study of Pleiades and Hyades star clusters, estimation of the distance to the Large Magellanic Cloud and measurement for orbits of binary stars.

His work was recognized by:

- Honorary degrees from Utrecht, Copenhagen, Paris
- Bruce Medal of Astronomical Society of Pacific
- Darwin Lectureship and Gold Medal of Royal Astronomical Society London
- Ole Romer medal awarded by city of Copenhagen

KARL SCHWARZSCHILD AND
EJNAR HERTZSPRUNG

Henry Norris Russell (1877-1957) was an American astronomer with studies at Princeton University New Jersey, U.S. and Cambridge University England.

He was named the *Decan of the American Astronomy* for his contribution to the exploration and research of the universe.

Starting with his initial work at Cambridge University on the determination of stellar distances, Russell began to assemble data from different classes of stars observing that these data relate spectral type and absolute brightness.

Russell knew that the stars with the same temperature (which from spectrum is stated) give the same quantity of radiation per square kilometre of their surface and searched to find a relation between the temperature and the brightness of a star.

Also when a weaker and a brighter star at equal distance from Earth show the same temperature then must be the weaker star smaller than the brighter star.

Independently of Hertzsprung but using similar methods, Russell determined the distance from Earth for more stars and by that he could determine their absolute brightness.

Soon he concluded that were actually two general types of stars: giants and dwarfs.

In 1910 year the astrophysicist Karl Schwarzschild introduced him to Ejnar Hertzsprung.

In 1913 year Russell also represented the result of his research graphically for stars whose spectral classes were known and which were not in clusters but so nearly that their distances and thus their absolute brightness could be determined.

He plotted the spectral classes of stars - sequence O B A F G K M from Harvard, which give also the colour and the temperature of the stars against their absolute magnitude.

His diagram also showed the distinction between the giants and main sequence stars like Hertzsprung's colour-magnitude diagram, looking like Hertzsprung's but rotated 90 degrees clockwise.

The two diagrams have the same topography and collectively are known as

Hertzsprung-Russell Diagram or **H-R Diagram**.

By an accident of history an apparently independent parallel study of the relation between brightness and spectral type in the Pleiades star cluster was undertaken by **Hans Rosenberg** of Gottingen whose overlooked paper contained the first published H-R diagram in June 1910.

Until the apparition of the H-R diagram, the most astronomers assumed that the stars development follows the spectral sequence, namely each star begins hot and blue-white and ends cool and red, considerations due to the classification system developed by Antonia Maury,

Annie Cannon and other researchers from Harvard.

HENRY NORRIS RUSSELL

Against, Russell proposed that both types of red stars giants and dwarfs, he and Hertzsprung found, represent the first respectively the last step in the life cycle of a star, each star beginning its presence as red giant of spectral class M, then contracts and heats itself, runs down the entire diagonal in H-R Diagram and ends as cold, red dwarf again in class M, conclusion known as premature.

Russell was engaged in research work including eclipsing binaries and detailed analysis of the spectrum of Sun. He postulated that most stars exhibit similar general combination of relative elemental abundances (dominated by hydrogen and helium) known as "Russell mixture".

Russell was an accomplished teacher and author of the *Solar System and Its Origin* and co-author of the *Astronomy,* excellent textbooks, which served as a guide for future research in astronomy and astrophysics. Russell received recognition from more organizations:

- Gold Medal - Royal Astronomical Society England
- Henry Draper Medal - US National Academy of Sciences
- Rumford Prize - American Academy of Arts and Sciences
- Bruce Gold Medal - Astronomical Society of Pacific

Hertzsprung–Russell Diagram is one of the most useful and powerful plots in astrophysics. Its usefulness comes from how it illustrates the many different types of stars in one glance and is remarkable in how it essentially creates a graphical way to represent the stellar complexities in one simple plot.

The Hertzsprung–Russell Diagram is a graph displaying the characteristics of any star.

It provides essential clues for the evolution of stars and has been a valuable tool in determining the distances of those stars beyond the reach of parallax measurements.

Accurately, the horizontal axis should display the effective temperature / colour / spectral classes of stars and the vertical axis should display the luminosity / brightness of stars.

Usually:

In surface temperature the range is

from 3000 K through 4000 K, 5500 K, 7000 K, 10000 K, 18000 K to 40000 K

In colour the range is from bluest through bluish, blue-white, white, yellow, orange to red

In spectral classes the range is the seven letters **O B A F G K M** easy remembered by the phrase: **O**h, **B**e **A F**ine **G**uy/**G**irl: **K**iss **M**e

In luminosity the range is $(10^{-5}...10^{6})$ L$_{sun}$

In absolute brightness the total range is 27 M [(+19...–8)M]

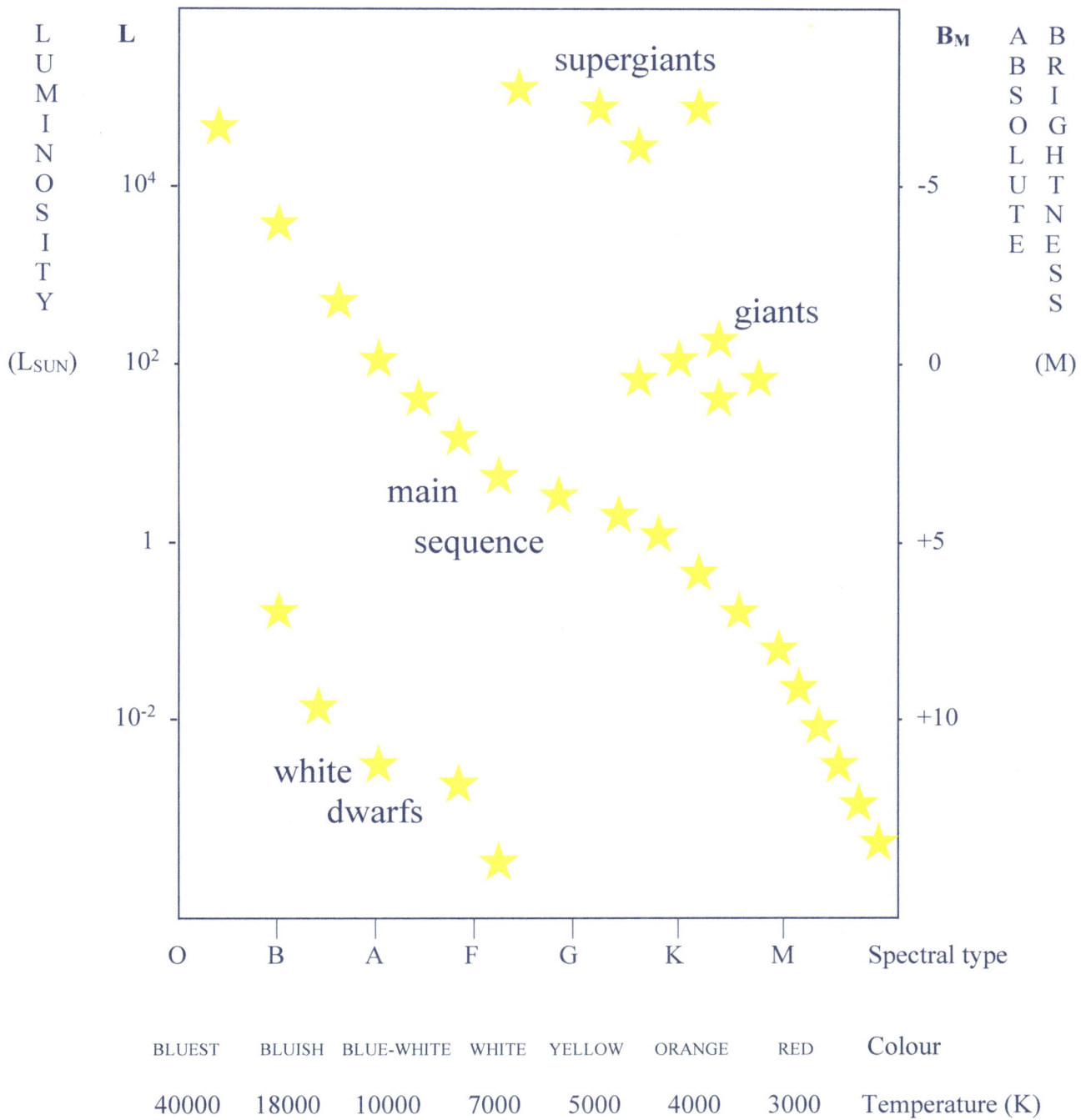

HERTZSPRUNG – RUSSELL DIAGRAM

FIG 8

The brightest stars appear at top of the H-R diagram, the vertical axis having the most negative value of the brightness B at top.

The outcome is not a random scattering of points, the stars populating very specific regions of the diagram and that gives clues about their physical nature and their stage of evolution.

Equal of what stars are plotted in a H-R diagram, it shows that the stars are not distributed uniformly but concentrated manly in four regions, seen in FIG 8.

The four regions are:

1.   A band well populated (~90% of stars) running diagonally from high luminosity, hot surface temperatures to low luminosity, cool surface temperatures named *Main Sequence*

(most stars of Sun size)

2.   A very high luminosity region, scarcely populated at top of diagram with extremely large stars called *Supergiants*

3.   A region with stars of luminosities between those of the supergiants and main sequence stars termed *Giants* (stars of Earth orbit size)

4.   A region with stars, small but hot, termed *White Dwarfs* (stars of Earth size)

The modern theorists say that is hard to believe that the stars travel the main sequence, but they have a place depending on their mass. The mass decides also when a star leaves the main row.

The stars, which are rich in mass are also hot and very bright corresponding to the blue colour and the pattern of their spectral lines are the O and B stars. In their centre has place the fusion process of high delivering energy. In the course of their storm like development they use fast their matter, leave the main sequence and swell up in giants and supergiants building groups in the right upper corner of diagram. Massive stars can end their lives in *Supernova* explosions, which leave behind *Neutron Stars* or *Black Holes*; as they are detected at non-visible wavelengths they are not represented on the H-R diagram.

The yellow stars of middle mass like our Sun belong to the middle part of the main sequence, are colder and less bright then O and B stars, consume their matter slower and remain milliards of years on the main sequence before they swell up in giants, finally contracting in white dwarfs.

The stars which remain on the main sequence, are cold, weak lighting M stars with small mass, which use their fuel sparingly, their estimated life spans longer than the estimated age of the universe.

The H-R Diagram version from FIG 9 illustrates that for the stars on the *main sequence*, the diameter hangs together with the temperature and the luminosity.

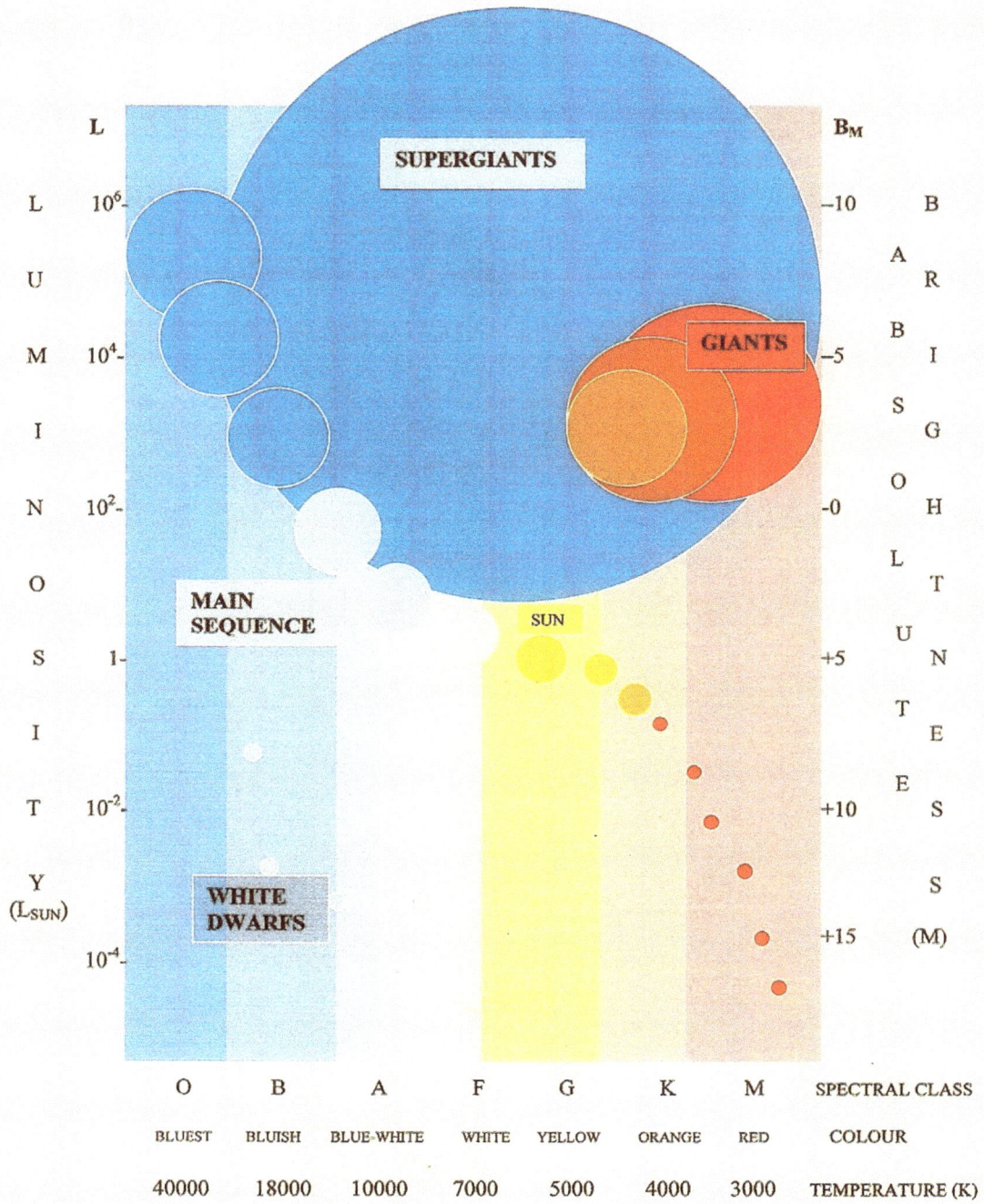

HERTZSPRUNG - RUSSELL DIAGRAM

FIG 9

The hot, luminous, blue stars upper-left are also the biggest; the massive, bright, yellow stars like our Sun are accordingly smaller; and the cool, faint, red stars are the smallest bodies on the sequence called *dwarfs*; the statistics show that stars fainter than Sun are far more numerous than those brighter than Sun. However a minority of stars *off* the main sequence, has high temperature but small diameter and low brightness; they are the *white dwarfs*, in general hotter than Sun but having a small surface, light much weaker; vice versa are red *giants* and many *supergiants* mostly not hotter than Sun but millions times brighter because their diameter rises up to 1000 times above that of Sun.

In FIG 10 is presented a H-R Diagram for 20 brightest stars in the sky and a few fainter, nearby stars (such as Barnard's star and Kapteyn's star).

The Properties of the *Stars of Main Sequence* include:

The mass μ of the main sequence stars varies from 0.08 μ_{sun} to 200 μ_{sun}.

Correspondingly the radius R of the main sequence stars varies between 0.1 - 25 R_{sun}.

The temperature of the main sequence stars varies from 3000 K to 40,000 K.

The main lines in the spectrum of the stars on the main sequence show that the atmosphere of stars contain neutral metals like Ca and Fe, ionized Ca, neutral H and He, ionized He.

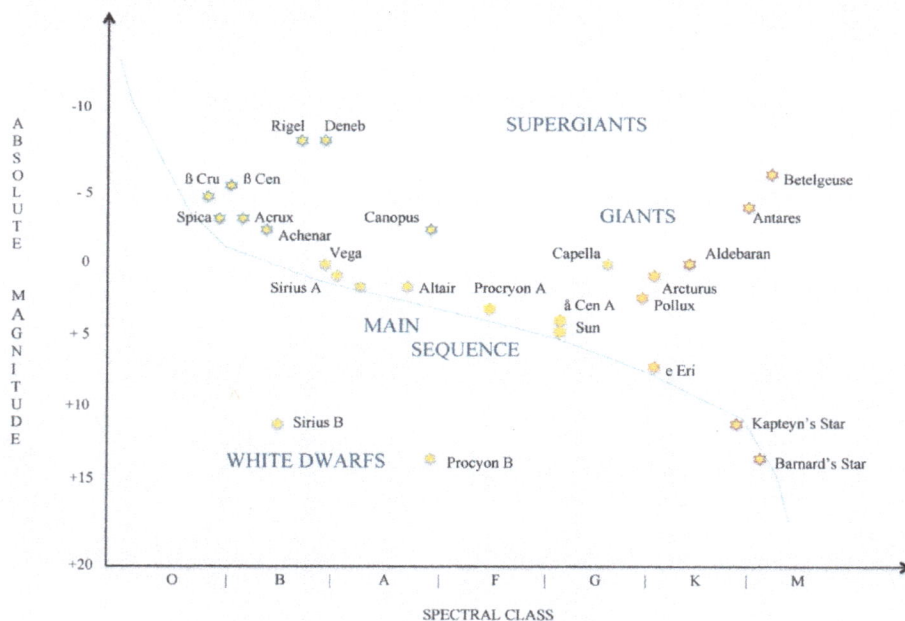

H-R DIAGRAM FOR BRIGHTEST STARS OF SKY and few nearby

FIG 10

112

references

REISE DURCH DAS UNIVERSUM - VON DER REDAKTION DER TIME-LIFE BUCHER AMSTERDAM 1989

HISTORY OF ASTRONOMY - AN ENCYCLOPEDIA - JOHN LANKFORD 1997

ENCYCLOPEDIA OF ASTRONOMY AND ASTROPHYSICS - PAUL MURDIN 2001

THE ILLUSTRATED ENCYCLOPEDIA OF THE UNIVERSE - IAN RIDPATH 2001

VOYAGES THROUGH THE UNIVERSE - ANDREW FRAKNOI, DAVID MORRISON,
SIDNEY WOLF 2001

A TO Z OF SCIENTISTS IN SPACE AND ASTRONOMY - DEBORAH TODD, JOSEPH ANGELO JR
2005

THE FACTA ON FILE-DICTIONARY OF ASTRONOMY - JOHN DAINTITH, WILLIAM GOULD
2006

BIOGRAPHICAL ENCYCLOPEDIA OF ASTRONOMERS - THOMAS HOCKEY 2007

PHILIP'S ASTRONOMY ENCYCLOPEDIA - PATRICK MOORE 2011

DATA BOOK OF ASTRONOMY - PATRICK MOORE 2014

THE PHYSICS FACTBOOK

THE ENCYCLOPEDIA OF SCIENCE

ASTRONOMY AUSTRALIA

THE SAO ENCYCLOPEDIA OF ASTRONOMY

MEDIA RELEASE INFORMATION - CSIRO

WIKIPEDIA

WIKIMEDIA COMMONS

ANNEX

Original file (1,500 × 1,865 pixels, file size: 418 KB, MIME type: image/jpeg);

Deutsch: Joseph von Fraunhofer war ein deutscher Optiker und Physiker.
English: Joseph von Fraunhofer was a German physicist.

Quelle: Engraving in the Small Portraits collection, History of Science Collections, University of Oklahoma Libraries.

File:Robert Bunsen 02.jpg

Original file (1,732 × 2,227 pixels, file size: 1.72 MB, MIME type: image/jpeg);

Open in Media ViewerConfiguration

Description	**English:** Robert Bunsen (* 30. März 1811 in Göttingen; † 16. August 1899 in Heidelberg)
Date	Unknown
Source	http://ihm.nlm.nih.gov/images/B03904
Author	C. H. Jeens

File:Gustav Robert Kirchhoff.jpg

Original file (815 × 1,151 pixels, file size: 366 KB, MIME type: image/jpeg)

Gustav Robert Kirchhoff Source

Licensing

File:Angelo Secchi.jpg

From Wikimedia Commons, the free media repository

Original file (835 × 1,148 pixels, file size: 590 KB, MIME type: image/jpeg)

Pietro Angelo Secchi (1818-1878), Italian astronomer.

Licensing

File:PSM V22 D302 Henry Draper.jpg

From Wikimedia Commons, the free media repository

Size of this preview: 541 × 599 pixels. Other resolutions: 217 × 240 pixels | 433 × 480 pixels | 693 × 768 pixels | 924 × 1,024 pixels | 1,887 × 2,091 pixels.

Original file (1,887 × 2,091 pixels, file size: 969 KB, MIME type: image/jpeg); ZoomViewer

File:Edward Charles Pickering 1880s.jpg

From Wikimedia Commons, the free media repository

Original file (742 × 987 pixels, file size: 437 KB, MIME type: image/jpeg)

Description	Edward Charles Pickering
Date	1880s
Author	Unknown (Mondadori Publishers)

Permission
(Reusing this
file)

This work is in the **public domain** in the United States because it was published (or registered with the U.S. Copyright Office) before January 1, 1923.

⚠ Public domain works must be out of copyright in both the United States and in the source country of the work in order to be hosted on the Commons. If the work is not a U.S. work, the file **must** have an additional copyright tag indicating the copyright status in the source country.

File:Astronomer Edward Charles Pickering's Harvard computers.jpg

From Wikimedia Commons, the free media repository

Original file (1,920 × 1,502 pixels, file size: 591 KB, MIME type: image/jpeg);

Description	**English:** "Pickering's Harem," so-called, for the group of women computers at the Harvard College Observatory, who worked for the astronomer Edward Charles Pickering. The group included Harvard computer and astronomer Henrietta Swan Leavitt (1868–1921), Annie Jump Cannon (1863–1941), Williamina Fleming (1857–1911), and Antonia Maury (1866–1952).
Date	circa 1890
Author	Harvard College Observatory

Licensing

File:Williamina Paton Stevens Fleming circa 1890s.jpg

Original file (1,481 × 1,920 pixels, file size: 515 KB, MIME type: image/jpeg);

Description	**English:** Williamina Paton Stevens Fleming (1857-1911), circa 1890s. (Courtesy Curator of Astronomical Photographs at Harvard College Observatory.)
Date	circa 1890s
Author	Unknown

File:Antonia maury.jpg

From Wikimedia Commons, the free media repository

Antonia_maury.jpg (200 × 298 pixels, file size: 24 KB, MIME type: image/jpeg)

Description	**English:** Photograph of Antonia Maury from her senior year at Vassar College
Date	circa 1887
Author	Vassar College

Licensing

File:Annie Jump Cannon 1922 Portrait.jpg

Original file (888 × 1,250 pixels, file size: 183 KB, MIME type: image/jpeg)

Description	**English:** Mrs. Annie Jump Cannon, head-and-shoulders portrait, left profile. Library of Congress permalink.
Date	1922
Author	New York World-Telegram and the Sun Newspaper

File:Portrait of Henry Norris Russell.jpg

Original file (741 × 981 pixels, file size: 236 KB, MIME type: image/jpeg)

Description	**English:** Portrait of Henry Norris Russell
Date	15 October 2013, 01:09:50
Author	Unknown

Licensing

File:Karl Schwarzschild and Ejnar Hertzsprung (1909).jpg

Description	**English:** Karl Schwarzschild (left) and Ejnar Hertzsprung in professorial gowns in front of the Göttingen Observatory building sometime in 1909.
Date	1909
Author	The Göttingen Observatory photo, courtesy Hartmut Grosser.

Licensing

www.ingramcontent.com/pod-product-compliance
Lightning Source LLC
Chambersburg PA
CBHW041154220326
41598CB00045B/7425